ON GENES, GODS AND TYRANTS

CAMILO J. CELA-CONDE

Dean of the Faculty of Arts, University of Palma, Mallorca

ON GENES, GODS AND TYRANTS

The Biological Causation of Morality

Translated by Penelope Lock

D. REIDEL PUBLISHING COMPANY

A MEMBER OF THE KLUWER ACADEMIC PUBLISHERS GROUP

DORDRECHT / BOSTON / LANCASTER / TOKYO

Library of Congress Cataloging-in-Publication Data

Cela-Conde, Camilo Jose.
    On genes, gods, and tyrants.

    Translation of: De genes, dioses y tiranos.
    Bibliography: p.
    Includes index.
        1. Ethics.   2.Biology—Moral and ethical aspects.   I.   Title.
BJ58.C4513 1987   171'.7   87–16409
ISBN 1–55608–024–7
ISBN 1–55608–036–0 (pbk.)

Originally published in Spanish under the title:
De genes, dioses y tiranos
by Alianza Editorial, Madrid, 1985

Published by D. Reidel Publishing Company,
P.O. Box 17, 3300 AA Dordrecht, Holland.

Sold and distributed in the U.S.A. and Canada
by Kluwer Academic Publishers,
101 Philip Drive, Norwell, MA 02061, U.S.A.

In all other countries, sold and distributed
by Kluwer Academic Publishers Group,
P.O. Box 322, 3300 AH Dordrecht, Holland

To Francisco J. Ayala

*Neither in gods, kings nor in courts
is the Supreme Saviour*

The International

TABLE OF CONTENTS

FOREWORD

> Our future was with the collective, but our
> survival was with the individual, and the
> paradox was killing us everyday.

> John Le Carré *Smiley's People* (1979)

Since the time of Ancient Greek lyrical poetry, it has been one of man's dreams to explain his own conduct. This is the background to all his activities, from literature to speculative philosophy, including those odds and ends which, for want of a better name and more precise boundaries are called "human science". Over the past nine or ten years a new member has been added to this inquisitive family, one which, moreover, claims to be scientific to an extremely high degree: biology. This is in fact a recurrent event, since theses designed to introduce causal biological explanations into the general field of human action had already been formulated on at least two occasions (in original Darwinism and the Neo-Darwinist synthesis).

Ethologists and sociobiologists are today taking over and assuring us that they have the necessary tools to provide an answer to what perhaps seemed the most slippery subject in the hands of science: the social being. As might be expected, philosophers have reacted with some scepticism. Though human conduct is undoubtedly subject to determinants, the lion's share of responsibility lies with society itself. At the time when biology was beginning to develop the theories necessary to overcome creationism, Karl Marx had already managed to construct highly sophisticated interpretive models of human social behaviour. Even those who take pride in questioning the ability of Marxism (or indeed of any other set of theories,) to construct "laws of history" – something they consider impossible – are to a certain extent indebted to Marx because their criticism makes use of the body of doctrine he introduced. Marxism may be of no use in

predicting our actions, but it is impossible to understand human
social behaviour without bearing in mind a large part of Marx's
teachings.

The biological contribution is therefore faced with the tragic
risk of being true in the long run, but irrelevant. Its determinants
may come to the surface only at the cost of forgetting those others
from the social field which disguise and transform primary biological impulses. Ethologists and sociobiologists would deny this,
and would even reverse the terms, but the difficulties they meet in
finding support for their case are still so immense as to be
discouraging.

However, there is one aspect of human behaviour where the
presence of biological determinants may be more balanced in
relation to social determinants: moral action. Ethical behaviour
is, in the final instance, a matter of individual decision; a characteristic which favours biological theses. Unfortunately, this is also
where free will comes in, that is to say, human will proud of its
bill of freedom and reluctant to admit determinants of any sort.
But this may be a mirage like so many others created from an
anthropocentric point of view. So it would be somewhat risky to
reject biological determinants in morals by resorting to the *petitio
principii* which by definition identifies ethics with freedom.

The following pages are an attempt to establish the relationship
between biology and morals, taking as a starting-point certain
theses related to Darwinism. The aspect of social determinism
will hardly be touched on, a decision which, in my opinion, is
sufficiently justified by the overwhelming amount of literature on
the subject. This does not imply, however, that the role of society
is irrelevant as a source of determinants, and its presence will be
felt in some cases. After all, our future lies in the collective. But
what is under discussion here is not which is the greater source of
moral determinants, assuming these exist, but rather, to what
extent we can accept a biological basis for the ethical phenomenon.

The informed reader will no doubt be aware of a blatant
absence: that of Frankfortians and Erlangians, who have probably gone most deeply into the ethical and political aspects of
linguistic universals. I take the blame for this, since my intention
is to justify a second part to this book dealing with the problems
of legitimizing collective instances by making use of many of the

theoretical constructs to be raised here. Should anyone be misled
by such cryptic words, I suggest he look up the paper I presented
at the *III Jornadas de Etica y Filosofía política*, held in Granada
in 1983, entitled "La legitimación política de la ética" (The
political legitimization of ethics), where I make a formal statement of my intentions.

I should like to clear up a few other points. Throughout the
text, the words "morals" and "ethics" will be used as equivalents.
This may seem an unfortunate choice, since "morals" has recently been used only for normative morals and "ethics" for
metamorals in this sense. But I hope to be granted this licence as I
am going to propose a different classification of the various
aspects which intervene in the phenomenon of moral action. In
any case, I may quote the illustrious precedent set by Ferrater
Mora (1981, p. 19), who chooses the same solution. Ferrater is
one of the rare philosophers in Spain to have dealt with sociobiological matters and I owe a lot more to his book (written jointly
with Priscilla Cohn) than just the licence to identify ethics and
morals. The specialist reader may also be taken aback by my lack
of precision when referring to norms, criteria, judgements, laws
and values, without making more accurate distinctions. Or by my
reference to only one sense of such evaluative terms as "good",
when so much paper has been devoted to monographs attempting
to establish the meaning of such a word. It is a question of
understanding that biology projects extremely primary determinants, when at all, on the moral field, and under no circumstances
are these open to such sophistications. I hope that those prepared
to draw more accurate boundaries for themselves will find in my
proposals sufficient means by which to do so.

I also use "biological determinant" and "genetic determinant"
indistinctly. I should, indeed, distinguish between them (as anthropologists in the style of Malinowski have done), but insofar as the
determinants to be discussed here (those proposed by sociobiology) clearly refer to genetic content, I allow myself this liberty[1].

The manuscript was sent for their inspection and criticism to
several people, who helped me to straighten out my ideas and
weed out some nonsense here and there. I cannot claim to have
exorcised all my mistakes, nor can I refer to all their suggestions
one by one. For their patience and good intentions I wish to

thank Francisco Ayala, Gustavo Bueno, José Ferrater Mora, Manuel Garrido, Ernesto Garzón Valdés, Enrique López Castellón, Emilio Lledó, Jacobo Muñoz, Alberto Saoner and Amelia Valcárcel. I also wish to publicly express my gratitude to the Ministry of Culture for granting me an aid to literary creation in order to write the book, and to the Universitat de les Illes Balears for the aid I was granted for translation. My gratitude also to Cathy Sweeny and Alberto Saoner for reviewing the English translation. And, finally, I am most grateful to Francisco Ayala, Michael Ruse and Edward O. Wilson for their help in promoting the English version.

In the following pages I have made use of material from some earlier articles of mine, more or less modified in all cases. These articles are: "Una aproximación a la 'Hipótesis de las ideas innatas' de Noam Chomsky" (*Mayurqa*, 16, 1976, pp. 139–188); "La virtud del azar" (*Taula*, 1, 1982, pp. 7–13); "En torno al concepto de simpatía" (jointly with Alberto Saoner. Actas de las las Jornadas de Etica e Historia de la ética, Madrid U.N.E.D., 1979, in press -sic-); "Tres tesis falaces de la ideología liberal" (*Sistema*, 50–51, 1983, pp. 51–60); "Nature and Reason in the Darwinian Theory of Moral Sense" (*History and Philosophy of the Life Sciences*, 6, 1984, pp. 3–24); and "La determinación ética según los modelos r y K de la genética de poblaciones" (Actas del II Congreso de teoria y metodología de la ciencia, Oviedo, 1983, Vol. II, pp. 377–380).

NOTE

[1] I wish to thank José Sanmartín for pointing out this possible abuse during the sessions of the international symposium on the philosophy of Karl Popper (Madrid, November 1984)

CHAPTER 1

MORAL LEVELS

> When T. H. Huxley states that ethical
> sense is not a product of, but is rather
> opposed to, biological evolution, he is con-
> sidering principally whether moral codes
> generally accepted by humanity are in-
> spired by biological evolution and at the
> same time promote it; although T. H.
> Huxley's terminology is at times ambi-
> guous, it appears that he does not intend to
> deny that ethical capacity is rooted in bio-
> logical nature. J. S. Huxley and C. D.
> Waddington support this root and also
> claim that biological evolution specifies
> those moral codes to be accepted and pre-
> determines mankind to accept them;
> neither of these two authors seems to real-
> ise clearly that biological determinism of
> ethical capacity does not necessarily imply
> a determinism of ethical standards to be
> followed. A distinction between ethical ca-
> pacity and ethical standards is found,
> though not always explicitly, in the work of
> Dobzhansky and Simpson.

> Francisco J. Ayala: *Origen y evolución del hombre* (1980)

This book is concerned with moral questions, considered almost always from a biological perspective. Just reading the above quotation from Ayala reveals that this task is neither easy nor free from confusion. In fact, it is not usually even considered by philosophers to be worthy of attention. Until recently, those engaged in the study of morals would limit the scope of genuinely philosophical ethics to the analysis of the language in which moral preference was expressed.[1] This cannot be said to be the standard approach to ethical questions since the now distant times of the

1

early Greek Enlightenment. Nowadays we can concern ourselves
with such sophisticated matters as whether the sentence 'Why
should I be moral?' has any sense beyond mere tautology. By
contrast, throughout the history of ethics moral philosophers
have tended to consider the whyfores and wherefores of this
question, either by explaining why I *am* a being with moral
preoccupations, or by discussing how far those who tell me I
should do a given thing are right.

Both approaches come under what could in a broad sense be
called the (philosophical) matter of moral conduct. The parenth-
eses here are not merely ornamental. The first approach would
lead to a sort of science or technology of conduct, to a discipline
linked methodologically to the need to offer causal explanations
(though this proves impossible with our current resources). This
is the opinion widely held among philosophers, or at least among
those who come under the protective mantle of analytics. Only
the second merits the honour of being considered a philosophical,
that is, speculative approach. The following chapters will deal
with the origin and scope of this dual outlook; it is enough for the
moment just to mention its existence.

In the mid-seventies moral philosophy underwent one of those
sudden changes of course which periodically lead the interest of
specialists in surprisingly divergent directions. If the reduction
made by the analytic school had one clear virtue it was that it
banished once and for all the normative side of morality, which
had too often been linked to religious or lay moralistics, from the
interest of philosophers. It comes as a shock that the conjunction
of what in the eyes of philosophers are somewhat exotic disciplines – ethology, population genetics, entomology, molecular
genetics – should have created a new field for ethical discussion
stimulated by the decidedly polemic intentions of scientists fond
of speculative extrapolations. This field is far removed from the
analysis of moral language and closer to the classical questions of
normative ethics.

This book aims to offer some opinions about classical problems
in moral philosophy, supported on occasions by the theories of
biology. This means therefore taking up the gauntlet thrown
down by the natural sciences. It is necessary to explain how the
challenge is accepted. Leaving aside the deontological leaning of

biologists, the relationship between biology and ethics is somewhat stormy and demands that a radical choice be made between
the following statements: (a) biology can offer explanations of
ethical problems which are relatively superior to those offered by
any other discipline, including philosophy; and (b) biology is not
related in any way to what ethical problems are really about.

It would be simple to offer a list of authors who favour one
statement or the other. However, I believe that the most interesting
ideas about the relationship between ethics and biology, or
if preferred between philosophy in general and biology, belong to
those authors who could not be classified by such a dichotomy.
These are people who in general accept the basic biological
assumptions and Darwinian hypotheses in particular, but who
consider the extrapolation offered by sociobiologists to be false.
To a large extent the entangled thread of the biological challenge
has begun to be unravelled thanks to their help. My recognition
of this commendable work does not mean that I am going to
follow a similar strategy, that is, of setting out in detail the
possible fallacies contained in *On Human Nature* or *Genes, Mind
and Culture*[2] and contrasting them with the more traditional lines
of thought. Neither do I intend to summarize the theses of Mary
Midgley, Peter Singer or Michael Ruse, to extract something like
a 'general view' of the question. This is not because I believe the
task to be unworthy, but essentially because I consider that there
is a basic mistake in all these intelligent criticisms of sociobiology
(a mistake to the extent that my critique proves correct). The
error lies in *accepting* the biologists' viewpoint when referring to
the phenomenon of 'the moral'.[3] Perhaps the complexity of the
theoretical artifice constructed by biology forces us to ignore the
complexity of human ethical activity itself. I consider that such
servitude leads to the whole question of the relationships between
ethics and biology being wrongly framed from the outset. Sociobiology identifies man's moral behaviour almost exclusively with
the phenomenon of altruism. This is understandable. Population
genetics has developed powerful theses about how the individuals
of a reasonably gregarious species can improve their adaptation
to the environment and their chances of survival through the
course of altruistic behaviour.[4] Even though the Darwinian principle of survival in individual terms is still valid, in my opinion, in

tenets of natural selection, there is a claim for the existence of
behaviour strategies capable of increasing adaptive success through
acts which may initially seem harmful to the being acting in an
'altruistic' fashion by reducing his own chances of survival, albeit
in a limited and subtle way, in favour of the chances of other
individuals in his group. The key to understanding such behaviour, paradoxically connected to natural selection, lies in the
concept of *inclusive fitness*: altruistic behaviour by the members
of a group supposedly benefits the whole group, which means that
the balance for each member of the group is greater than it would
be through strictly individual adaptive strategies.[5]

Given that the very mechanism of natural selection seems to
allow for a type of conduct opposed in principle to individual
selfishness, it is not surprising that sociobiologists have seized
upon the phenomenon of altruism as the foundation upon which
to build their bridge between ethics and biology. Consequently it
does not seem strange that those who seek to criticize determinist
theses have stressed parallel analyses of altruistic behaviour. It is
immediately obvious what is peculiar about Wilson's concept of
altruism which he considers to be when a person or animal
increases the fitness of another at the expense of his own fitness.
Things are not as simple as Wilson implies and not even biologists
are willing to accept without discussion that any activity of this
kind is an act of altruism. Imagine for example that a father dives
into the water to save his son from drowning, putting his own life
at serious risk. For Robert Trivers (1971) this would not be a
genuinely altruistic act, since from a biological perspective it
could be interpreted as contributing to the rescue of the father's
own genes present also in his endangered son. Trivers defines
reciprocal altruism as that behaviour capable of benefiting an
individual who is so far removed in genetic terms that it could not
be explained by evolution through kin selection. According to
Trivers' model, reciprocal altruism could even benefit an individual of another species (1971, p. 214).[6] At this point we are not
concerned with the explanatory power of the models of altruism
advanced by Trivers, Wilson, or Alexander (1977), or of any
other involved in the biological polemic on inclusive fitness. We
need to appreciate that we cannot take the theories of sociobiological altruism and apply them mechanically to the problem

of ethical altruism because we are dealing to a large extent with a linguistic distortion. B. C. R. Bertram has tried to argue that the biological definition of altruism should be similar to that used in everyday human language. Yet even his own definition of biological altruism:

behaviour which is likely to increase the reproductive output of another member of the same species who is not a descendant of the actor, and which at least in the short term is likely also to reduce the number of the actor's own descendants (1982, p. 252)

hardly fulfils that condition. Human altruism, as a person's moral behaviour is understood in certain circles, need have no direct bearing on reproductive output, not even in a subtle way unknown to the agents. Any scientist involved in the search for theories to explain conduct which is hard at first glance to justify from the perspective of natural selection is aware of this fact, and this is precisely why one talks of *biological* altruism as opposed to *moral* altruism. I do not intend to confuse these terms, but I do question whether the best definition of biological altruism need be the one closest to the sense of moral altruism as Bertram claims. I can see no logical connection, unless it is first established that the two phenomena are in fact related, which is extremely doubtful. After all, we are not trying to give moral and biological altruism a different name and definition, but to realise that they are two different types of behaviour and only by analogy is the term 'altruism' applied to both.

When Trivers tries to offer a complete model of reciprocal altruism (biological of course), he has to indicate the conditions necessary to prevent 'cheats' from taking advantage of the benefits offered by a group of altruistic individuals. The advantages of reciprocal altruism lie in the fact that the cost of losing adaptive fitness through altruistic behaviour are sufficiently compensated, statistically speaking, for the individual who finds himself in a situation where he needs help. It is obvious that in a population of altruists, the strategy of a selfish mutant would be perfect; he would receive all the benefits without risking anything in exchange, and this essentially is the argument used by Dawkins in defence of the selfish nature of genes in the metaphorical species pool (1976, Chapter 5). Trivers assumes a group of altruists could avoid the drawbacks of having these cheats by behaving beneficially only towards those who themselves show altruistic tendencies. The problem is that although formally it is easy to define this strategy and do the mathematical calculations of cost and benefit with regard to the chance of biological selection, it is hard to imagine *how* the individuals of a species practising altruism could manage to spot, deal with, and prevent the existence of fraudulent egoists. It is an unhelpful tautology to say that altruistic behaviour will be selected whenever the cost/benefit balance is favourable. We are interested in knowing what specific type of behaviour an altruistic individual will develop when he must decide whether to offer help to a fellow-being in danger. Games theory has developed a model of analysis of alternatives between altruistic and selfish behaviour which has become famous under the name of 'Prisoner's Dilemma' and which Trivers quotes in support of his arguments. The hypothetical situation is of two prisoners in solitary confinement who are successively submitted to an interrogation in which each in turn is offered his freedom in exchange for a confession. If one confesses he is set free, but the other receives a long-term prison sentence. If neither confesses, they will continue to be interrogated for some time, but in the end both will be set free. If *both* confess then both receive a medium-term sentence, longer than the time appointed for the interrogation, but less than the sentence one of them would receive if the other confessed. The interesting question is which strategy is the best.

Peter Singer has examined the Prisoner's Dilemma in detail in relation to the sociobiological theses on altruism (1981, pp. 145 ff.) and has reached the rather obvious conclusion that the best strategy of all is that adopted by those who obey altruistic motivations no matter what. However, he adds a remark with which I cannot so readily agree: "The Prisoner's Dilemma explains why there could be an evolutionary advantage in being genuinely altruistic instead of making reciprocal exchanges on the basis of calculated self-interest" (*ibid.*, p. 47). If this is all the Prisoner's Dilemma tells us we are lost. What really concerns us is how an altruistic group can defend itself from the presence of selfish

individuals willing to cheat whenever necessary. Singer is limited
to the paths of pure tautology. He imagines a situation in earlier
phylogenetic times when two primitive men are attacked by a
sabretooth tiger, a more believable situation in evolutionary
terms than the Prisoner's Dilemma to be sure. He puts forward
the best strategy as that of the altruistic hunters, even though he
concedes that "Of course, an egoist who could find an altruist to
go hunting with him would do better still; but altruists who could
not detect – and refuse to assist – purely self-interested partners
would be selected against" (ibid., pp. 48–49). I maintain that to
select those individuals who could do so leads us nowhere. Unless
we can explain the means used to detect and decide, and the
benefits of doing so in order that these means are genetically
selected in the process of hominization, the models from games
theory will only tell us that this altruism is all very well, but it is
fundamentally an unviable strategy.[7]

There are two ways of understanding how the existence of
altruism can be assured: through strong innate conditioning (such
as in social insects) or through sophisticated means of evaluating
conduct and decision-making. Sociobiologists usually refer to the
latter when talking of human altruism and its connections with
animal altruism. Certain social mammals show altruistic behaviour in the biological sense. We talk of altruism as an ethically
valuable quality, and thus the continuist thesis is established: the
presumably certain origin of biological altruism can lead to certain obvious conclusions about moral altruism. But this formula is
mistaken because continuity can in no way be taken for granted.
In fact, altruistic behaviour may not even be considered the
fundamental element in the whole ethical process. It seems to me
that through continuist theses an error is persistently committed,
one which Darwin had considered to be dispelled when he brought
together his own theory about the extent of the biological determinism of human morals, which will be dealt with in the next
chapter. Let us just anticipate a little now with an episode which
illustrates the sense of what I propose, which is to consider animal
altruism, analysable in terms of population genetics, as a phenomenon of functional adaptation analogous, but not homologous to
human moral actions.[8] It seems certain that man can count on a

measure of genetically determined altruism, such as that found in
the animal world, but this is in no way the key element upon
which to base the phenomenon of ethics. I shall try to explain
why.

When Darwin speaks of moral feelings in Chapter IV of *The
Descent of Man* he mentions the apparently heroic behaviour of a
group of baboons in the case of the rescue of a baby baboon from
the clutches of predators at the risk to the life of the brave adult.[9]
Darwin does not regard this as a moral action, and goes on to put
forward his own thesis about *moral sense* which he deals with
later.

Why aren't there ethics among baboons? If we take altruism to
be a fundamental element in ethical behaviour it would not be
difficult to accept. Sociobiology does in fact admit homology in
this sense: Wilson makes a parallel analysis of the characteristics
shared by human and termite societies and includes ethics in
both. Termites sacrifice themselves altruistically for the good of
the community (which was a problem which certainly caused
Darwin concern: how can sterile individuals transmit their altruistic make-up to the next generation?). If we reduce human ethics
to a core of altruistic behaviour (in the sociobiological sense)
there cannot of course be much objection to Wilson's comparison.[10] Personally I believe there is a basic error of consideration
about ethical phenomena. Of course, 'ethics' includes altruistic
conduct, but in a different sense than in population genetics.
There is no close relationship however between 'altruistic conduct' and 'moral conduct', there is no dependent relationship
which would allow us to reduce moral conduct to altruism and
proclaim therefore that all behaviour which is altruistic is morally
acceptable. This is a derivation which seems to be linked to the
existence of altruistic conduct which can be empirically confirmed
from the field of biology. Unfortunately, in no way can such
conduct be automatically considered as ethical conduct.

'Ethics' is closely linked to the existence of criteria of appraisal,
that is, to communities capable of expressing such judgements in
their language. Any supposedly altruistic behaviour can be dis-
counted ethically through *ad hoc* criteria, a fact empiricism discovered early on. When Adam Smith laid the foundation for an
ethical rationalism which ended up forming a substratum for

utilitarianism, he was careful to warn against the acritical application of a mechanical theory of sympathetic identification. We do not endorse conduct as would a well-designed machine through a system of sympathetic links; we analyse the situation to check whether the conduct (altruistic or whatever) comes up to accepted standards. In this way altruism is undoubtedly related to moral behaviour in the broad sense of the word, but ethical choice can in no way be reduced to the altruistic act. An ethical decision can be understood in fact as something *disconnected* from altruistic behaviour, as happens in the Kantian thesis of the ethical illegitimacy of the lie – we are obliged to hand over a defenceless woman to her probable murderer rather than lie and declare that she has not taken refuge in our house. Such an act would be repugnant to any consistent altruist, but we cannot discard the Kantian viewpoint which considers that we are faced with a morally reprehensible or even 'unethical' idea on these grounds alone. As I try to establish here, the connection between ethics and altruism is more complex and necessarily implies the construction of criteria capable of modifying more than a little any likely innate tendency to establish and maintain altruistic links between members of a population. I shall return to this matter later.

It is evident that the appearance of elements such as the moral standard, judgements on situations, comparison of criteria and the like complicates matters and places us in a field where the history of moral thought has already wasted countless words. But to ignore the existence of these problems is not in my view an acceptable alternative. If altruistic conduct alone is no guarantee of moral conduct and it proves necessary to add complicated operations of analysis, we have no option but to do so. I think the whole analysis of moral phenomena is enriched by the contribution of contemporary biology, but not so much that by ignoring classical problems they can be considered solved.

Biology can throw some light on the question of altruism, and even on the thorny problem of heroes, but no light at all will be shed unless we know what we are talking about when we speak of the relationship between ethics and biology. The statement could be put in the most general way:

There are ethical phenomena which are determined by biological phenomena

So the thesis I support is one which denies the feasibility of an immediate discussion of the validity or invalidity of that statement. Perhaps we are sufficiently well-informed about what may be understood as 'determining biological phenomena', although for my part I believe that not even the attempts of Lumsden and Wilson (1981) to offer translucent box models of genetic programming have succeeded. I am however totally convinced that the expression 'ethical phenomena' is excessively vague and needs to be further defined if we are to discuss the determination of morals by the genetic substratum.

What does Wilson's oft-quoted claim about temporarily removing ethics from the hands of philosophers and biologizing it actually mean? What does biologizing ethics mean? Wilson does not explain in *Sociobiology* and even less in his philosophical proposals in *On Human Nature*. The mechanistic perspective of *Genes, Mind and Culture* does not do much to clarify matters, because to remove an ethical problem (altruism or any other) from the hands of philosophers and biologize it can be understood in very different ways. It may be claimed simply that biology can explain why ethical impulses exist in man, a proposal which philosophers unwilling to admit any kind of biological determinism would have to agree with. Or it may be assumed that biology can sustain ethical criteria of preference, which would signal the end of the thorny problem of rational preference. Or it may be considered that, thanks to biology, humanity can firmly express which ultimate values are worthwhile, and normative ethics would live a new and final splendour. All these ways of understanding the biologizing of ethics meet the conditions necessary to be considered part of the bridge which might span the two disciplines. The danger comes if we confuse them. The various meanings of biological determinism can be so wide that the expression 'to biologize ethics' can take on almost opposite senses depending on how it is understood. It we hold a discussion in *all* these fields (which rarely happens) we would have to concede in the end that 'ethics' and 'biology' are neither closely related nor totally distinct. We might discover that strong determining factors do exist as far as the genetic weight of ethical impulses is concerned. We might in fact draw any number of partial conclusions since 'ethics' is not a single and indivisible entity which has to be considered in its entirety for what is being said to be understood. On the contrary, there are distinct aspects within this and other ethical shades of meaning which will force us to make different analyses when trying to trace possible proofs of biological determinism. It is meaningless to state categorically that such determinism does or does not exist in a general sense, because there is nothing like a general sense in the field of ethics. In fact, the separation between different levels of a supposedly unique essential ethical phenom-enon has become a commonplace within the specialized litera-ture. Reference is usually made to Kant and his 'Copernican revolution' when referring to the distinction between the motive and criterion of a moral action. I do not intend to detract from Kant (indeed, the strict duality he proposes deserves much more detailed attention), but I think the most solid antecedents for such a proposal are to be found in the empiricist tradition. The thread leading from Adam Smith to Charles Darwin, taking in the nineteenth century moral sense philosophers, leads us to the clearest formulation of the relationship between ethics and bi-ology as far as the foundations of the 'motive to act' are con-cerned.

The distinction between motive and criterion is basic if we are to get rid of a lot of confusions about the biological foundation of moral phenomena. But in my opinion this is not enough. Today sociobiology claims, for instance, the existence of ecological determinants capable of making certain values, related to the r and K adaptive strategies studied by population genetics, appear in human groups. This means going beyond theses about the biological origin of moral conduct or ethical preference. It means that techniques of the natural sciences can empirically detect dependence on the genetic substratum within codes present in societies. It is no longer a question of psychological foundations, nor of a tendency towards certain ways of resolving conflicts. It means the appearance of strong connections in the strictest sense which natural law could have ever imagined.

To delineate the field of discussion more clearly, I propose to start from four different levels, each one of which is related to distinct aspects of the 'moral phenomenon', in order to trace possible determinants whatever type they might be:

1. The alpha-moral level, which contains those matters related to the moral character of the human being, the possible replies to questions about why man is a moral being. The existence or not of instinctive tendencies towards altruism and egoism, the presence of elements capable of classifying conduct as 'moral', and everything which relates to feelings, emotions, character, impulses, and psychological mechanisms in the sense indicated, fall under this category.
2. The beta-moral level, which deals with the criteria man uses to classify an action as morally desirable. The process of argument, both in structure and content (excluding certain aspects related to the following levels) the meaning of evaluative words, the presence of moral judgements and their possible validity, comparison of codes, etc., are elements which enter into this type of discussion.
3. The gamma-moral level, which comprises empirical normative rules and moral codes within groups, apart of course from their eventual justification or discussion.
4. The delta-moral level which consists of the question of ultimate ends of a moral character.[11]

None of the four levels can be completely reduced to the others in any existing system, although there are obviously close relationships between them. The separation of the alpha and beta levels has been a commonplace since the Enlightenment, as I said earlier. The gamma level, for its part, clearly enjoys the differential character it gains by its crystallization into empirical codes which can be detected and classified. Perhaps the delta-moral level has the most need to justify its separate existence. It may be argued that it could be conceptually reduced to the former three, and in fact there are ethical systems which deny the privileged character of ultimate ends and others which simply leave the question to one side. But as long as there are propositions about ultimate ends which go beyond doubting them or regarding them as relative to a positive code, (closely dependent for example on divine beings or natural forces) it seems best to set them in a level of their own in order to make discussion clearer.

Perhaps an example will make the sense of the separation I propose between the levels more evident. Let us suppose that a citizen is sleeping on the terrace of his home when he is woken by shouts from the street. He looks out and sees down on the pavement two men struggling with a woman, apparently trying to rape her. Imagine the moral situation the alarmed and now wide-awake neighbour finds himself in. He may be a heartless fellow and unaffected by the fact of the woman being left to her fate, or perhaps he is a timid, easily-frightened person, terrified as much by the idea of going down and facing the evil-doers as of calling the police and getting involved as a witness in a possible court case. Or perhaps he is simply paralysed by fear. But he might also be a good citizen, capable of sacrificing his comfort and safety to the sense of justice he feels and is ready both to call the police and to run down and help the victim. If we consider such a range of possibilities and analyse what happens, we approach this action from the alpha-moral level.

But, suppose the subject who realizes that the rape is happening is not alone. His wife, or a friend, is also on the terrace and expresses his or her opinion, not about the character and willingness of the person hesitating between offering his assistance or not, but about the extent of one's moral duty to help a victim of attempted rape, and the circumstances in which this becomes an obligation. The reasons put forward belong to the second level, the beta-moral level, unless the citizen in question ends the ethical discussion at once by admitting that moral duty to help a woman who is being raped exists, but that in his case he considers himself a coward and a blackguard and pays no attention to such ethereal matters.

But the question does not end here. Throughout the discussion, if time permits, references may be made to norms prevailing in the society to which all these people belong. Perhaps the right thing to do is to go straight to the aid of any neighbour in difficulty; perhaps there is even a law about it. Perhaps we may even imagine that among some more or less remote peoples the custom is to join in with the act of rape. All these questions refer

to codes existing beyond those being discussed, to elements
included in the gamma-moral level, even when the authority of
the code will no doubt be relevant in guiding the criteria on which
decisions are based. Even when used as an argument for moral
action, these elements unquestionably have a meaning, an empirical body which lends them a certain objective character. An
anthropologist would obviously distinguish between cultures con-
demning the rape, cultures which remain indifferent, and cultures
which react violently on behalf of the victim.

If the alarmed neighbour and his guest are not what we might
call men of action, and are of a somewhat Socratic disposition,
the discussion about existing norms and which criterion to apply
could very well go on and on, with one reason after another being
questioned. Even so, a final point would be reached as soon as
one of the participants used an argument such as 'we must go
down and help that poor woman because she is a human being
and the human condition demands solidarity' or 'we must stay
here because it is a sin to interfere in the course of fate'. These
statements may also be questioned of course, but if they are
accepted by all the participants moral action has been referred to
a special type of norm, to an ultimate end which we classify as
delta-moral. The above-mentioned anthropologist may attempt
to intervene by talking about the *etic* as opposed to the *emic*, but
since I shall deal with this matter later we will keep him quiet for
the moment. Even if ultimate ends are a part of the cultural
heritage of a group, the transcendental meaning given to them by
those who define and accept such supreme values justifies reserving them a final level.

All the levels I propose are so closely related that it may prove
difficult to distinguish clearly what refers to one of them in
particular. It is all the more necessary then to explain the different meanings if one wants to avoid serious confusion in the
treatment of moral phenomena. Richard D. Alexander for instance refers to the process of endoculturation in which the
notion of 'right' and 'wrong' is acquired. He claims that:

Parents begin instilling the ideas of right and wrong in their children, and this is
probably the normal origin of the concepts for most individuals. Initially, at
least, right and wrong are defined for children as whatever their parents say is

right and wrong. What, though, are usual concepts of right and wrong in parents' views of their children's behaviour? One might suppose that children are simply taught by their parents never to deceive, always to tell the truth, the whole truth, and nothing but the truth; therefore, that children are taught always to be altruistic toward others, to be certain that justice is afforded to all those with whom they interact, and that their own interests are secondary to those of others or of the members of the group to which they belong.

Alas, it cannot be true. (1979, p. 274).

What concerns Alexander is that the eventual behaviour of children does not allow us to infer that this is the way in which they gain access to 'right' and 'wrong', and that in fact parents teach their children to deceive and pretend in such a way as to maximize their conduct in relation to the 'inclusive fitness' of the individuals within the group (pp. 174–175). I am not now going to enter into whether this may or may not be accepted as a feasible model of endoculturation with regard to moral values. The determinism which can be attributed to the models of population genetics and inclusive fitness as a central concept in socio-biology will be dealt with in the relevant chapter. I should just like to point out that Alexander is quite simply confusing the beta and gamma-moral levels. What children pick up as the concept of 'right' is a result of their biological condition and, as I shall try to argue in the chapter devoted to the beta-moral level, of something else as well. But the fact that in a given group it is considered right to adopt altruistic behaviour towards older relations, or to tell the truth, or to deceive in some cases, is a gamma-moral problem related to the particular moral patterns followed within in, or a delta-moral one if it refers to principles such as the truth.

The confusion is logical. In a significant and ultimately individual way, a moral deed is successively connected to the person who acts and to the one who judges – who may of course be the same person. When it becomes generalized (characteristically in the gamma-moral level) it is always provisional and subject to the need to recover individual intimacy through being assimilated by the person who must make use of such norms and criteria, that is, in the process of endoculturation. In such an individualized phenomenon, it is not easy to distinguish between those elements of, say, a structural nature (beta-moral) and their empirical manifestation (gamma-moral). All the more reason, therefore, that I insist that we must distinguish carefully between the different levels.

The individual level is not the only source of relationships. The levels of moral phenomena are also interrelated through the possible influence they might have on each other. It is obvious that the type of ethical criterion maintained in the beta-moral level will make certain positive norms possible in the gamma-moral level and will prevent others. The type of ultimate end detected in the delta-moral level will also impose obligations on the others. These more obvious connections will hardly be considered here as I aim to focus on the much more neglected question of the relative independence between various aspects of 'morals'. I shall develop my idea by investigating the relevance of biological determinants which may be projected onto each of the central levels which have obsessed supposedly rational ethical systems. Clearly this limits me to a non-reductionist theory of ethics in the sense of one open, at least partly, to causal explanations. To speak of the existence of causal explanations in the field of ethics is anything but original; I assume then that I may stay within the field of biological determinants and leave aside almost everything related to sociological determinants and their own, relevant, causal explanations. There is already too much literature dedicated to these for it to be of any use to add more. Nevertheless, I will make certain exceptions which refer to the appearance of those cultural, or if one prefers, social determinants which are closely linked to biological ones in the formulations advanced by sociobiology.

CHAPTER 2

THE ALPHA-MORAL LEVEL: IN THE BEGINNING WAS DARWIN

> No distinction is more usual in all systems of ethics than that betwixt *natural abilities* and *moral virtues*; where the former are placed on the same footing with bodily endowments, and are supposed to have no merit or moral worth annexed to them. Whoever considers the matter accurately, will find, that a dispute upon this head would be merely a dispute of words, and that, though these qualities are not altogether of the same kind, yet they agree in the most material circumstances.

> David Hume, *A Treatise of Human Nature*

As I said earlier, when the different levels of moral behaviour are submitted to analysis, at the present time even those philosophers most reluctant to accept any type of biological determinism have no choice but to concede, at least, that the existence of ethical impulses in human beings is something which could possibly be explained in terms of genetic evolution. This concession usually comes grudgingly, since the good intentions of classical empiricism have in general been ignored by those who have stuck to the sharp dualism established through the distinction between science and philosophy.

In fact the disinterest of philosophers, or rather of some philosophers, in biology and psychology has been increasing as responses from these disciplines to questions about human nature have been shedding more light on our behaviour. It is as though Wittengenstein's plea to be silent on those matters about which we know nothing had been taken to mean the opposite. The leap implied for biology by Darwinism as a theoretical body could not help but approximate the criteria for, and in the long run introduce an at times only implicit consensus on the foundations of moral behaviour. This type of consensus can be illustrated by reference to an author so openly hostile to scientific thinking on these matters that he called the chapter from which the following idea is taken 'Ethics without Biology': there Thomas Nagel states that biology can tell us something about the perceptive and motivational beginnings of ethics, but he considers that in its current state it has little relation with the thought processes through which these starting points are transcended (1979, p. 146).

Anything is better than nothing. It seems then that we can take the idea of biological determinism at the alpha-moral level as paradigmatically accepted and go on to discuss the second part of Nagel's paragraph. But before doing so, I should like to put forward some ideas on how such a thesis appears in the field of biology (within the theory of ethics constructed by Charles Darwin in his mature period) and in what sense biology is heir to the empirical thinking which dates back to Hume and Smith.

The philosophical significance of Darwinism in general can be illustrated by a commonplace: the collapse of rationalist optimism throughout the 19th Century as a result of the work of the great figures of the new Irrationalism. It was Darwin who was in large part responsible for the connection made between rational criteria and nature. Regardless of whether it was Schopenhauer or Paul Ree who inspired the following words, Nietzsche applauded the sense of the idea with jubilation, and not a little irony: "How did reason come into the world?" – we read in *Morgenrothe* – "In a rational way, as it should be: by chance. This chance will have to be solved as an enigma". According to a generalized interpretation which I am not going to argue with, the solution to the enigma in Darwin acquires the form of a causal dependency.

Nevertheless, it would be a risk to dismiss the specific relationship between nature and morals in Darwinian ethical postulates without going any further. However much Darwin continues to support the biological origin of everything within the realm of morality, (the content of the four levels I pointed out earlier) he can in no way be held up as the champion of the irrationalist hierarchy. I shall explain the development of the Darwinian theory of ethics by means of the following theses:

1. Darwin constructs a theory of ethics which is related to the authors of the moral sense school. In it, an explicit distinction is maintained between alpha and beta-moral levels, between *motive* and *criterion*. Even so, certain links do appear.
2. The connection between the two levels is made in a sense completely opposed to the current proposals of the biologization of ethics, that is, with the alpha-moral level being dependent on the others.
3. The rationalist hierarchy can be defended from Darwin's position thanks to Lamarckian interpretations of the process of genetic inheritance.
4. The meaning of Darwinian rationalism was anticipated by those authors (Hume and Adam Smith) who inverted the classical theory of moral sense.

In Chapter III of *The Descent of Man*,[1] Darwin openly sets out to demonstrate that there is no essential difference between the faculties of man and those of other superior mammals. The latter have simple faculties – memory, attention, association of ideas, and imagination, all of which Darwin analyses and explains – and in consequence they also, in a broad sense, have rationality, since the characteristics of abstraction and self-consciousness which we associate with rationality are nothing more than a combination of the simple faculties shared by all. But it is the next chapter, Chapter IV, which contains the central body of the Darwinian theory of ethics,[2] which takes the form of speculation about the similarities and differences between man and other gregarious species. Group formation is something natural to all, and the fact of association can be explained through the existence of social instincts. But, in the community life of man a new phenomenon appears: moral sense; something capable of profoundly changing certain aspects of social conduct, of introducing distinguishing features, something, in short, related to an instinctive element present in man on which ethical conduct can in principle be founded: sympathy.

Darwin draws on a solid tradition in the school of Scottish moralists for his analysis of sympathy as the phenomenon responsible for social life. In the course of a discussion, which in its origins dates from Shaftesbury and Hutcheson, one thing at least becomes patently clear: that the sympathetic mechanism cannot be put forward simultaneously as the motive and criterion of moral action. Adam Smith's confusion and ambiguity in this respect have led to critiques of his work, not only from Thomas Brown (*Lectures on the Philosophy of the Human Mind*, 1820, with express mention of sympathy), but from James Mackintosh (*Dissertation on the Progress of Ethical Philosophy*, 1836, who directs his attacks against William Paley), John Austin (*The Province of Jurisprudence Determined*, 1832) or Samuel Bailey (*Letters on the Philosophy of the Human Mind, Third Series*, 1855–1863). These all agree on the need to distinguish the 'moral faculty' from the 'standard', even though their own positions on moral sense in no way overlap.[3]

The context in which Darwin constructs his own theory of moral sense includes acceptance of Kantian dualism. This would allow the possibility of dealing only with moral aspects, to the neglect of ethical criteria, thereby entering into one of the great controversies of the moment: whether or not there exists an innate moral sense on which human actions can be based. In spite of the general tendency to be expected from the work of a naturalist, Darwin goes on to put forward strong theses, not only on beta-moral criteria, but on gamma-level norms of empirical morality, and even on questions related to ultimate ends which I consider delta-moral. A similar combination can already be found in the analysis of the sympathetic mechanism made by Alexander Bain (*Mental and Moral Sciences*, 1868) whom Darwin draws on when he talks of moral sense. Bain considers sympathetic contact to be capable of shaping the feelings and points of view of others, in such a way that creeds, feelings, and opinions achieve a strength able to justify their uniformity and conservativeness.[4] It is Darwin who offers an integrated model of the process and its phylogenetic significance.

Moral sense for Darwin is the sum of social instincts and intellectual faculties among which figures critical rationalization. In the long run they all function in harmony and compatibility, so that, in this scheme, moral actions are based on instinct, assisted 'in a natural way' by reason and experience. This is not a simultaneous process. From a phylogenetic point of view, moral sense is acquired through an evolution of four stages which are set out at the beginning of Chapter IV of *The Descent of Man*. Even once the faculty has been acquired we would still be at a very primitive stage of morality. There would be very different motives for action, including social instincts and appreciation of the opinion of others, as well as a certain degree of selfishness. As time passes, there would be a tendency to leave selfishness as a motive for action aside and give greater value to the judgements and values of our fellows.

Presented in this form, the model meets with a serious difficulty. If, as was indicated earlier, Darwin is easily able to distinguish between the first two moral levels, between motive and criterion, why now does he posit a clearly beta-moral element – the opinion of others – as a basis for conduct?

The answer lies in the specific idea of the development of moral sense adopted by Darwinian theory. Since it can be explained in diachronic terms, three steps must be taken into account:

(a) The presence of sympathetic instincts and some primary moral notions, together with the role played by personal egoism, result in action which is probably influenced strongly by selfish interest.
(b) but the reflective nature of man's intellectual faculties forces him to reconsider the meaning and results of this primary action, so that
(c) a firm resolution is taken to act in a different, that is, altruistic way in the future.

Since this final resolution means the reinforcement of the sympathetic instincts present in human nature, the general sense of the process is the increase of moral content in successive actions, the gradual increase in the relative weight of sympathy and consequent loss of selfish interest. As we shall see later on, the assimilation of such 'outside criteria' (the element which gives the model coherence) is explicable thanks to Lamarckian interpretations of inheritance. Before going any further, let me emphasize the significance of the connection made between the alpha and beta-moral levels.

The Descent of Man is full of arguments and examples intended to strengthen the idea of an evolutionary continuity between man and the other superior mammals, but Darwin considers that "the difference between the mind of the lowest man and that of the highest animal is immense" (p. 494), a fact he attributes to morals. At times however it is difficult to accept this idea. Let us look at an example. Making use of an account from the naturalist Brehm, Darwin comments on the behaviour of a group of baboons who are attacked by dogs at the foot of a mountain they are trying to climb. The cries of a young baboon, surrounded on an isolated rock, lead an adult male to abandon the safety of the higher peaks and come down to rescue the little one. Darwin considers this baboon a true hero, to use his words (p. 474), and there is no doubt that we would call such an act heroic if we substituted the word 'man' for 'baboon'. Why then do we not give moral value to this action?

The tradition of the moral sense school has a criterion capable of distinguishing between moral and non-moral actions, which is also applicable to this pseudo-heroic behaviour of baboons: the criterion of motivation. Actions carried out on impulse would not be moral, while "actions done deliberately, after a victory over opposing desires, or when prompted by some exalted motive" would (p. 482). But Darwin finds a solution of this type irrelevant as it leads to the slippery ground of which motives are to be considered exalted enough to provide the basis of moral conduct. So he immediately suggests another possible criterion, that of the *moral being*. All those actions observed in man and animals which seem heroic and could cause confusion will be called and considered 'moral' only when they are carried out by a being capable of comparing past and future motives and actions, and consequently approving them if warranted. Thus only man acts morally because only he possesses moral sense. Suspicions about circularity in this argument can be discarded because in the Darwinian theory of moral sense there is an element which specifically characterizes this human faculty, making it possible to provide something more than a mere *petitio principii* to establish differences between man and animals: critical rational conduct. From the moment human reflection intervenes as a differentiating factor, action is determined by two basic elements, both of which originate in the *alpha-moral* level; two biological principles on which the anthropological condition rests: instinctive sympathy and the capacity for critical reasoning. Neither of these is exclusive to man in qualitative terms, but only human beings possess these qualities in sufficient degree to achieve moral conduct.

Thus nature is able to endow man with the capacity to carry out actions which should be characterized as right or wrong. But the existence of empirical communities which, in a manner natural to all of them, have different criteria about what may or may not be considered morally acceptable, obliges us to enter the terrain of ethical preference. In my opinion the solution offered by Darwin fits perfectly into the empirical tradition which has already established the need to make the substratum of moral feelings compatible with the task of rationalizing and interpreting actions under judgement. Regarding the relationship between instinctive sympathy and rational criterion it contains assumptions similar in some ways to those used by David Hume, to whom I shall refer later. We must appreciate that the use of rationality as a characteristic of moral activity (even when relative to the criterion) is a delicate subject among those authors who base moral sense on the alpha-moral principle. With qualified exceptions, the general strategy of the empiricists is a firm opposition to the principle of basing morals on rational activity, thence the search for psychological mechanisms capable of adequately characterizing 'motives for action'. But, as we shall see later, this in no way implies a total abandonment of the rationalist perspective. When formulating his theory of moral sense, which I referred to earlier, Darwin could also rely on the distinction already common in his era between motive and criterion, thus allowing for rational intrusions. Let us see how these appear.

The characteristics of the sympathetic instinct alone oblige man to take account of the criteria of the fellows with whom he lives in close community. The desire to obtain the approbation of others obviously means accepting their opinions, and out of this consent appears a 'guide of conduct', certainly not a universal one, which in some cases – Darwin tells us – provokes "the strangest customs and superstitions, in complete opposition to the true welfare and happiness of mankind" and "have become all-powerful throughout the world" (p. 491). But we must take into account that for such an incisive sentence to be meaningful, it necessarily supposes:

(a) possession of a beta-moral criterion of preference capable of indicating which are the strange customs and superstitions among the huge number of different actions which we find in the empirical gamma-moral level;

(b) possession also of an ultimate delta-moral postulate which places the *summum bonum* in the "true welfare and happiness of mankind".

(c) finally, in an ethical system like Darwinism, the proposal of an interpretative scheme of the way criteria of preference and the ultimate desirable end are related.

According to Darwin, man has the means with which to distinguish superior moral rules from inferior ones and, of course, to classify certain customs as strange and superstitious. At the end of the chapter dedicated to moral sense, the golden rule is pointed to as the foundation of morals (in the beta-moral sense of course): "As ye would that men should do to you, do ye to them likewise", to which, with some help, social instincts lead. This does not seem to me to be sufficient to understand his position fully. What is really relevant is the mechanism of preference itself: a process of intellectual and rational improvement in which the consequences of actions are understood and "the knowledge necessary to reject baneful customs and superstitions" is acquired. This, then, is something which is concerned with rational progress linked to the growing task of reflecting on and considering alternatives.

The form in which ultimate ends are postulated is a matter which has led to the discovery of utilitarian contaminants in *The Descent of Man*, so much so that the suggestion made in the conclusions of the chapter on moral sense in the first edition of the work about the 'Greatest Happiness Principle' is completed from the second edition onwards with the specification that, in any case, this principle cannot be considered as the basis for moral conduct but as a criterion of sanction.[5] In a footnote, also in the second edition, Darwin insists that utilitarian causal explanations are inappropriate. All these specifications do not in fact seem necessary because the position of Darwin on ultimate criteria is quite clear already in the first edition of *The Descent of Man*. There, the utilitarian principle of the greatest happiness is substituted for that of 'the general good or wellbeing of the community', a concept coined from the interpretation of the development of social instincts in inferior animals, which in Darwin's opinion should be extended to include man. The general good is "the rearing of the greatest number of individuals in full vigour and health, with all their faculties perfect, under the conditions to which they are subjected" (p. 490). But this general good, or good of the community in the leap from animals to humans, meets certain difficulties of interpretation. Absolute instinctive determinism (which in the case of man would lead to having to regard all that refers to the general good as being part of alpha-moral nature in the strict sense) disappears, and in consequence we have the appearance of communities whose customs are evil and objectionable. Hence Darwin understands the necessity to limit the definition for reasons of political ethics, in a way which will become obvious when dealing with progress.

It may seem paradoxical to consider the modification of ultimate delta-moral ends (such as the good of the community) which by definition are desirable in themselves, from the criterion of beta-moral preference (critical reason applied to the sympathetic instinct). In fact this problem is ever-present when one tries to defend a theory of moral progress.[6] The paradox in *The Descent of Man* is resolved, in my judgement, thanks to Darwin's integration of all moral levels: of structural anthropology which is the foundation of the alpha-moral motive to act, the ethical criterion of beta-moral preference, the empirical existence of gamma-moral customs, and the delta-moral ultimate end of the general good of the community. The theory of natural selection leads in fact to a functionalist interpretation of morals which will be used explicitly in neo-Darwinism: the presence of ethics in man and in the human community is of such relevance to the adaptation to environmental conditions that both Darwin's original theory and the later proposals of the neo-Darwinists, ethologists, and sociobiologists have no alternative but to consider moral conduct as one of the main factors in the evolutionary stable strategies of the human species, regardless of whether the analysis itself of what these strategies are might differ.[7]

In fact it is Darwin who can maintain the closest link between all the levels I have distinguished in the phenomenon of ethics.

This is thanks to the way he interprets the internalization of ethical criteria by the individual and the existence of moral progress.

The system for explaining moral progress represents a return to the field of naturalistic irrationalism. This of course comes as a surprise. How can irrationalism appear in an area characterized by rationality as an anthropological guarantee? To understand this we must go back to the enigma remarked upon by Nietzsche, of chance capable of bringing reason into the world. Nature is in principle responsible for provoking the set of circumstances which resulted in the appearance of a superior mammal which has moral sense, and it will be nature once again, for Darwin, which makes moral progress something inevitable, almost automatic. But now it is a matter not of chance, but of a very Lamarckian necessity.

Darwin explains that in the course of evolution man gradually increases his level of virtue thanks to the intervention of inheritance, an intervention which is justified because increasing degrees of virtuous conduct become extremely positive elements in natural selection. But this does not imply a need for moral progress to be tied to inheritance. A functional theory of inheritance in the biological sense could alternatively explain how the moral character of man was written into his genetic code and in what way positive norms can mean certain advantages for a group's cultural evolution (for its progress in a sense which would not be strictly biological except in a sociobiological theory of culture). However, it is immensely difficult to marry both senses of evolution within today's working models. Darwin is able to do so thanks to Lamarckism, which inspired his analysis of how human communities progress. He believes that primitive man possesses distorted norms of positive morals which seek only the welfare of the group as a whole without taking any account of the interests of other men beyond it – despite the fact that man forms a single species – and of course, leaving aside all that represents the gratification of individual interests. With time this limited sympathy would advance, as civilization brings together growing numbers of citizens, in such a way that the truly civilized man would be he who finally understands, through the highest process of rationalization, that his sympathetic instincts should extend beyond the artificial borders of states and nations.

I think that this interpretative model of moral progress holds thanks to the fact it forms part of a comprehensive project in which what really progresses and evolves is a group of three parallel phenomena: the number of individuals in a group, intelligence and rational level, and the sympathy and benevolence of group members. This joint, growing, and harmonious evolution ensures advantages in terms of natural selection for those groups who are at once bigger, more intelligent, and more altruistic (even though altruism only functions within the community during the primary and intermediate phases) and these characteristics consequently go on increasing. What is extremely difficult to explain is how these apparently disparate phenomena can work together in such perfect harmony. It is inheritance of acquired characteristics which provides the elements necessary to balance the accounts. Darwin, quoting Spencer, expressed his conviction that norms of positive virtue would finally be incorporated into inheritance, so that the future of man would be oriented towards a peak where innate moral sense is capable of including not only instinctive sympathy, but also everything which human reason has considered ethically desirable and has incorporated into positive codes. This harmonious evolution of the alpha and beta-moral levels represents a great temptation for any empirical scientist. Given that the congruence is absolute, moral progress can be measured through a single indicator which cannot give rise to doubts: the size of the group. The more individuals a group contains, the higher will be its moral level and thus its rational capacity. In Social Darwinism this line of thought turns very quickly towards criteria of morality which are at the opposite end of the scale from altruism and benevolence, but although the roots of this operation may lie in Darwinian pangenesis, we can in no way assume that Darwin himself draws such conclusions.

In my opinion, even the panorama of moral progress is not enough to include Darwin in the rise of 19th Century irrationalism. His ethical naturalism is no more than an eschatological project, something which is directed towards an undoubtedly distant future where the hereditary accumulation of positive virtues will produce that man free from national ties and sympathetically linked to all his fellow beings. To all intents and purposes it is impossible to identify this comprehensive moral capacity with our own reality of human groups where sympathy still operates internally. Anyone who searches *The Descent of Man* for the dependence of moral criteria on nature will have to be content in our imperfect age with pointing out the existence of an innate sympathy which gives rise to centripetal activities in the heart of human groups. Effective moral conduct requires the addition of a process of reflection, and however much the capacity to reflect is, once more, natural to man, the result of this rational activity will not be linked to inheritance until the final Utopia is reached. Meanwhile, we will have to make do with the image of an innate sympathy linked to reflective processes on ethical judgements. I shall pause briefly to try to show the origin of this idea.

Through the development of various theories which come under the Scottish school of moral sense, sympathy is the concept which articulates the psychological explanation of the foundation of human morals. From the secondary and somewhat humble role it plays in the work of Shaftesbury, along with other synonyms which could vaguely be called 'social love', it gradually gains strength until it becomes the essential element in the theory of the basis of morals in Hume and Smith – with different manifestations of course.[8]

  In the early authors such as Shaftesbury and Hutcheson, the strategy of the Scottish thinkers of the moral sense school is to deny the ethics of selfishness, and even more fiercely of rationalism, as possible bases for man's moral conduct. Instead they dissect human emotions by empirical analyses and deductive demonstrations which postulate a theory of benevolent altruism as the motive power behind human actions. This form of interpreting what human moral conduct is represents without doubt the triumph of empirical psychology over those abstract rational principles which previously tried to characterize the foundation of ethics, but it would be dangerous to accept so readily that reason had been absolutely eradicated from the scene of the British Enlightenment. The fact is that within what in our perspective distinguishes the alpha and beta-moral levels, Scottish philosophers had discovered the value of empirical justifications for the former, for why there should be moral conduct among human

beings. And the achievements of this type of explanation con-
found the scene on the second level, the level of ethical preference
– though certainly not for long. Already in the sermons of Butler
there was a clear example of the attempt to reconcile rationalism
and sentimentalism; the great theories of moral sense constructed
by Hume and Smith offered ways of relating the two levels which
clearly anticipate the path which Darwin's solution will take.

As I said earlier, the concept of sympathy was going to play a
significant, but never a leading role in this line of thought. In the
context of enlightened British society, problems related to private
property and justice define the ground where the elements from
the moral sense school will meet. The discussion centres on the
natural law scope of rights and property. In contrast to those
justifications of private property based on its being a natural
element capable of balancing the forces of attraction and repul-
sion which are also found naturally in the relations between men,
and above all in the work process, (and which lead to mechanical
models explicitly compared to universal gravity in the work of
Hutcheson, or to animal nature in that of Lord Kames) Hume
tries to respond with a conventionalist explanation in which his
well-known thesis of artificial virtues appears.

Hume does not even concede that sentimentalism can provide
any profound explanation of man's conduct. For him, the means
of overcoming what is irregular or inconvenient about the affec-
tions as the ultimate motive for action will be an artifice. That
artifice is private property, which avoids a perpetual struggle for
the possession of goods (*A Treatise of Human Nature*, p. 489).
Hume is not attempting a history of property, but rather a
justification of its existence, and so needs to call upon what
establishes the property relationship between a person and an
object: justice. Through the analysis of justice he again repeats
the non-natural, but conventional character of the person-object
relationship and sets the origin of justice as an explanation for the
origin of private property. In fact it is the same artifice which
gives rise to both.

However, a more important question is pending: how can an
arbitrary convention function in such a reliable and general way
as to serve as the mechanism which keeps society from disinte-
grating? Hume's solution belongs to the psychological theories of

the moral sense school: feelings and impressions are the psychological vehicle which can offer a way towards the general acceptance of justice in human society. And specifically, sympathy is
responsible for the none too easy task of achieving the final
insertion of justice into the psychological plan, which, let us
remember, was an artifice. If selfish interest can lead to the
construction of a conventional means such as justice to ensure
social stability, sympathy for the public interest is the source of
moral approbation of just conduct (*ibid*., pp. 449–450).[8]

However complete the psychological explanation model of *A
Treatise of Human Nature* might be, there remain certain difficulties, particularly surrounding the ambiguous character of the
concept of sympathy in this text in its dependency on a theory of
limited benevolence. I have examined the problem more
thoroughly elsewhere,[9] and it is not worth insisting too much on
its consequences here, except as an indication of what we shall
find later in Darwin. The *Treatise*, primarily dedicated to the
analysis of affective elements of conduct, is not very concerned
with the crucial question of ethical preference. How could variations in legal codes be justified within a psychologistic model? It is
precisely the subject of the empirical diversity of codes and norms
which forces Hume at the end of the *Treatise* to introduce a
rational element of correction into his system of innate moral
sense. Hume explains that this sense always exists, but it takes on
new force through reflection. By reflecting on his condition, man
is able to extend his sympathy to all humanity instead of limiting
it to the results of the coming together of innate inclinations and
rare opportunities for personal contact. The memory of Darwin's
eschatological project inevitably looms up.

The most apparent elements of Darwin in the moral sense
school, if we agree to include Hume in this, appear in *An Enquiry
concerning the Principles of Morals*.[10] Here Hume offers a very
different psychological panorama. He begins by rejecting egoism
as the origin of moral sentiments and in consequence of justice
and property. It is social, not individual, interest which is now
significant, with which a new principle of moral sense appears,
oriented towards the ultimate end of the common good: useful-
ness. Although the extensive discussion of the psychological basis
of sentiments in the *Treatise* is not repeated, it is easy to appreci-

ate a radical change of direction: limited benevolence becomes an obstacle caused by the weakening of sympathy produced by physical distance; selfishness is a defect in the human character, and finally, reason is the element capable of correcting the affective disorders produced by these shortcomings. Through rational activity, man can overcome any barrier in the way of the extension of moral values and can establish general considerations of vice and virtue.

The leap from the artificial (primeval justice) to the natural (moral acceptance) depends once again on the mechanism of sympathy. But it can now receive a far more elegant treatment. Sympathy operates in the terrain of man's everyday relationships, enveloped in rational activity, as that which promotes contacts and ensures that the necessary corrections are made to the product of innate moral sentiment, to reach, finally, the ethical criterion of mature, civilized man. The model of the genesis of morality in man presented by Darwin is, as we saw earlier, extremely similar. Both coincide in presenting a final image of the triumph of consensus, benevolence, reason, and universality among all members of the human species. Both postulate very similar natural foundations to which they add a rational correction which is joined to what we now call the process of endoculturation. The differences of emphasis and approach are of course immense, but in what we could call the general strategy they coincide on the crucial points.

However, there is one detail of Darwin's ethical theory which should be seen in relation to an author already clearly distant from the proposals of the moral sense school: Adam Smith. This detail refers to the source of criteria for the moral acceptance or rejection of actions, that is, the question of the relation between those who act and those who observe and judge an action's morality.[11] Darwin also makes use of the audience as an element of contrast for moral judgements. Man adjusts his conduct to the criteria of approval of those who surround him, but not even in this way can a parallel be drawn between this formula of sanction and that of utilitarian theses which receive the element necessary for moral evaluation from Smith's 'maintained reason'. Utilitarianism certainly needs the presence of the *rational preferer*, since with Adam Smith's help it has got rid of everything related

to natural foundations of ethics which might previously have
existed. By contrast, Darwin turns to positions strategically close
to Hume, even though naturalist support is found in areas which
the moral sense philosophers never even suspected. The return to
innate moral sense in my opinion removes any trace of doubt and
forces us to reconsider the relationships between nature, reason,
and this time, evolution, from a perspective to which all ethical
philosophy, or if preferred, ethical biology, is today generally
committed.

Although the basic elements for an interpretation of how biological nature could determine human moral conduct are already
contained in the Darwinian theory of natural selection and the
evolution of species, the limitations introduced by the absence of
an adequate theory of the mechanism of inheritance impede
progress in the field of naturalist ethics. In fact, widespread
controversies produced towards the end of the century on the
theme of Darwinism and morals make more reference to ideological questions connected to Social Darwinism than to anything
else. It will be necessary to await the appearance of the neo-
Darwinist synthesis of Mendelian genetics and the theory of
natural selection, with the first black box models of the process of
character transmission, to rediscover the thread of naturalist
ethics.

The whole panorama of naturalist neo-Darwinist ethics is
governed by the idea of inheritance as a phenomenon of the
transmission of genetic information completed, in the case of the
human species, by the transmission of cultural information.[12] It is
a dualist perspective which postulates the existence of two different types of evolution for two different types of information
transmitted. But not at the cost of losing moral innateness which
appears indirectly and with greater pretensions than the modest
alpha-moral determinism which a few generations earlier satisfied
T. H. Huxley. However since beta, gamma, and delta determinants are going to be discussed in the light of sociobiological
positions, and these to a considerable extent modify the dual
neo-Darwinist thesis which separates nature and culture, I will
leave aside the dense contribution of Waddington, Dobzhansky
and Julian Huxley to the relation between ethics and biology. I

will confine myself to pointing out the nexus which allows us to sustain moral innateness, and in consequence, naturalism, inasmuch as this element does remain alive throughout the contributions of ethologists and sociobiologists. I refer to the functionalist theory of moral personality.

In 1960 Irving Hallowell published his *Evolution after Darwin*, a work which contains certain considerations on the role played by culture and society in the adaptation of the human species. In synthesis, the functionalist argument first regards individuals of different species as links in the transmission of genetic information. For individuals of the human species to fulfil this task, certain structures, foreign in themselves to the typical mechanism of transmission – the genetic code – are necessary: social and cultural structures. But these would not exist without preprequisites which are genetically determined, that is, a moral innateness which is translated in the phenotype by the tendency to use moral norms. In this way moral orders in the human being are functionally equivalent to other, non-moral, orders in the rest of the vertebrates: they are what make the task of the genetic transmission of characteristics possible (Waddington, 1960, p. 173 ff.).

The conclusion which Waddington draws from the functionalist theory of moral personality has two different aspects. The first consists of understanding that if social organization rests on innate moral mechanisms, genetic and cultural evolution will be compatible and in the ascendent. The second assumes the possibility of affirming the existence of a biological criterion of preference between opposing value systems. Neither of these two aspects has been free from criticism,[13] which in general refers to the presence of tautologous arguments and both naturalistic and genetic fallacies. I will not go into this question because the neo-Darwinist perspective is immediately superseded by the far more deterministic theses of the sociobiologists. However I should like to point out the essence of moral functionalism as an important element in explaining human phylogenesis and the incorporation of moral norms of conduct. This argument will be drawn upon with a certain frequency from now on.

CHAPTER 3

THE BETA-MORAL LEVEL: TO FEEL OR TO REASON. THE KANTIAN OBSTACLE

> It is doubtful whether these *why's* – the why's, in short, of the sciences explaining conduct – have anything to do with moral philosophy, as Kant wisely pointed out when he noted that however much the orders of *causality* and *rationality* might co-exist, they should not be confused. Let us suppose that at a given moment in my life I experience a sense of moral obligation, such as having to pick up a man injured on the road in my car, even if this means I may be late for an important appointment. If I want to know the cause of such an action – why I feel that I should pick up this injured man – a psychologist could tell me. (. . .) But if I wonder why I should pick up this man who I feel obliged to pick up, what I am asking about is in no way the cause of my feeling of obligation, but rather a reason for my duty.

> Javier Muguerza, *La razón sin esperanza* (1977)

Acceptance of the existence of a biological foundation for alpha-moral behaviour could after all be irrelevant for ethics. If Kantian dualism about causality and rationality is accepted (and there are reasons to do so) we find ourselves in a somewhat awkward situation, for reasons which arise from the analysis of the human process of knowledge.

In the *Critique of Pure Reason*, in the section which begins with the 'Solution of the Cosmological Idea of the Totality of the Deduction of Cosmical events from their Causes', Kant seeks to make the causality of nature compatible with the existence of freedom and human will capable of forming the basis for moral choice. Kant proposes two types of causality: one which comes from nature and one which derives from freedom, a distinction which allows him to talk of the existence of human will independently of the causality of phenomena.[1] Man's freedom implies the capacity to initiate a condition independent of all previous cause, which translated into ethical terms means the possibility of opting for some moral action without the intervention of determinants from the world of phenomenon capable of converting the supposed freedom into a fraud. Moral choice then, depends only on reason, but at the cost of converting *phenomenic causality* (which makes man as a phenomenon experience moral feelings) and *rationality* (which allows for ethics) into two separate orders. Kant is emphatic on this point (in relation to the principles of knowledge, reason constitutes a completely separate unity). And because of such a postulate, an ethical system in which duty and nature are related is established (the act to which duty is applied must necessarily be possible under natural conditions) and yet they are foreign to each other (the 'ought' when we consider merely the course of nature has neither application nor meaning. The question of what ought to happen in nature is as absurd as the question of what the properties of a circle ought to be. All we can ask is what takes place in nature, or in the latter case, what are the properties of a circle).

There are, then, three possible positions regarding the difficulties of accounting for the relation between nature and rationality as revealed in Transcendental Aesthetics. In the first place we can accept the consequences of Kantian metaphysics, maintain the dual perspective and reduce the field of ethics to speculations about rationality separate from the world of causalities. This in fact has been the strategy followed by the analytic school of moral language, with somewhat uncertain results as far as rational preference is concerned. For reasons which will become patent later when this matter is dealt with in Chapter 5, the task of rational preference seems to need better causal props than those allowed to it under the prism of Kantian dualism.

We could also opt to clear away ethical autonomy, proclaiming simply that Kant's metaphysics is mistaken. Now, where could the mistake lie? Perhaps in the idea of freedom as the capacity to initiate independent states. The second hypothesis would thus be to construct a metaphysical concept of liberty different from the Kantian (and, in fact, explicitly rejected by Kant)[2] of the type implicitly proposed by Wilson in *On Human Nature*, when he talks of the problem of free will. To initiate a new sequence free of prior causalities would be in itself quite simply an illusion: there are deep determinants, biological in origin, which impede such a move. But this illusion is impossible to demonstrate publicly and to predict because the determinist mechanisms are so complex that it is beyond the realms of possibility to anticipate human choice at a sufficiently detailed level. In *Genes, Mind and Culture*, a model of how such determinism works is even proposed, the influence of which I shall try to evaluate later. Through this measure the difficulty posed by Kant can be avoided, perhaps in a way which is somewhat cruel for the image man holds of himself, but we are so used to losing prerogatives about our human condition that one more need not make any difference, even if it refers to our introspectively sacred liberty. What is worse, at least for the proposition which inspires these pages, is that ethics such as derive from Wilson's metaphysical proposals are totally incompatible with claims of rationality. Equally, biological determinism extends to the supposed autonomy of reason.

Is it then metaphysically impossible to have a system of ethics which can claim simultaneously to recognize causal determinants and maintain rational criteria of preference, even if there remain insoluble problems in extracting empirical programmes of behaviour? I intend to deny this on the basis of a third proposal which consists of invoking a model of human knowledge capable of building a bridge between the two orders of causality and rationality. It may seem logically contradictory to claim to find something which is at once causally determined and yet is not, but this is only a semantic ambiguity. In maintaining that the order of rationality depends causally on the order of nature we are in fact doing nothing more than rescuing the old empirical postulate which appears in the final formulation of moral sense made by Darwin. Nature is responsible for the presence of ethics (and, let it be said in passing, for the reason claimed by Nietzsche) in the world. It is evident that this is not enough to explain matters, since to be content with invoking such a founding principle would be equivalent to making use of the genetic fallacy. However, this natural origin includes the maintenance of structural determinants of knowledge capable of introducing elements which can be causally traced in the beta-moral process of ethical preference. This is something quite different. In terms of Kantian metaphysics it is equivalent to maintaining the presence of synthetic a priori judgements, that is, knowledge prior to empirical activities which is at the same time significant for the fact of preference according to rational criteria. Should something like this exist, a gigantic step would have been taken towards the ethical system we are trying to found here, because the bridge between nature and reason would be built on a firm basis. A synthetic *a priori* judgement oriented towards ethics would be something belonging to the beta-moral level which could also be causally explicable. Victory cannot be claimed by relying on a circular argument: until now the only thing that has been achieved is to insist that it is necessary to find what is missing. To seek synthetic a priori judgements in moral matters is not at all easy.

Paradoxically one possible finding can be detected in the categories of Kantian Transcendental Aesthetics, as understood by the psychology of the last century in its interpretation of the principles of Space and Time as specific properties of the human mind.[3] In fact, the innatisms invoked by ethologists and sociobiologists are not far from a strategy of this type, although the appearance of cognitive psychology has complicated the scene considerably; rarely, with the exception of Konrad Lorenz, is reference made to Kant and the transcendental categories of knowledge. Unfortunately that is a programme which is of no use to us: it is full of innate judgements, but, as I said earlier, at the cost of a determinism so rigid that it invalidates claims of rationality. The task of tracing synthetic a priori judgements demands a different model of knowledge; a model like the one proposed by generative-transformational linguistics together with biology which finds inspiration in the Chomskyan school.

I am not going to repeat here the arguments used by Chomsky to support his famous "hypothesis of innate ideas",[4] but I should like to recall the reduction to the absurd from which he begins to postulate the black box model of the acquisition of linguistic competence. Data in the linguistic and extra-linguistic environment in which the child finds itself are so few, confusing, and disperse that if they were taken as the input fed into a cibernetic mechanism, which was to produce as output human language with its creative character, either the process would have to be considered miraculous or the computer in question credited with a structure sufficiently complex to compensate for the disproportions. "Innate ideas" form that structure of the human mind, genetically determined, which leads by certain programmed routes to the process of learning. This hypothesis has its correlates in the work of Lenneberg on the biological theory of the development of language. It is of no great importance that the nebulous black box models might affect the chances of validating Chomsky's hypothesis; what is of interest is the level of explanation it offers, and the fact that it implies a crucial consequence for the clarification of possible beta-moral determinants: what Chomsky and Lenneberg have done is to unite with strong links the epistemological and psychological levels. If Chomsky has, moreover, taken care to point out explicitly the risk run by transferring his hypothesis about innatism to fields other than that of language, I believe it is primarily for ideological motives.[5] The intimate connection between language and moral judgements in the ontogenetic process, demonstrated by cognitive psychology [6] would make it much more difficult still to explain the way in which these two areas could be separated. I hope this will be clear in the next chapter.

I stress that beneath Chomsky's claims of militant rationalism he has made a connection with empirical epistemology which is not to be scorned. The Scottish Enlightenment had already tried to find a psychological sense for the origin of knowledge which ended by intimately linking questions which were initially epistemological and questions which were finally moral, through the naturalistic ethics of moral sense. Guided by Hutcheson, Lord Home, and Shaftesbury we arrive at Hume, undecided between the universal sentiment of friendship and the principle of the *tabula rasa*; it is Kant who applies the rationalist knife and separates those things which are to do with knowledge and respond to criteria of causality, and those very different things which refer to its justification and respond to criteria of rationality. Now Chomsky presents a strong base capable of upholding once more the bridge between psychological and epistemological elements. It is true that the Scottish School of Moral Sense posed ethical problems; Chomsky, psychological problems; and Kant, epistemological problems. But the mechanism of language learning postulated by Chomsky, should it contain elements of ethical significance, that is elements related to the beta-moral level, becomes a source of innate determinants capable of affecting the origin of ethical motivations as much as moral justification itself through the link between psychological structure and knowledge. Still pending is an investigation of what this innate significant presence means for the beta-moral level, and on what biological basis it can be sustained. In the next chapter this matter will be touched on. I do not wish to close this chapter without saying something about the second of the propositions about the relation between the order of nature and that of rationality to which I referred at the beginning of this chapter: that which led to Wilson's metaphysical proposition of a limited concept of liberty.

The drawback of a determinist model such as Lumsden and Wilson offer in *Genes, Mind and Culture* (GMC from here onwards) is not that it is damaging to the pride and dignity of human beings, nor that it serves to sustain politically reprehensible policies such as racism, nor even that it presents insoluble problems in maintaining rationally based ethics. If human nature were such that it was painfully oriented towards racism and against rational solutions in the field of practical politics, it could be that our hopes for a world which was at least not worse would be seriously affected, but the validity of Lumsden and Wilson's model would not suffer from this (except that the authors as members of the human race might feel distressed at such a perspective). From my point of view the principal defect of the model is that it has gaps which threaten to invalidate it.

I am going to raise some points from reading Lumsden and Wilson's text, related above all to the first proposition of linking nature and reason. The motive for such bias will be understood if it is borne in mind that GMC is a book inspired by such a grandiose project that it includes areas of knowledge somewhat distant from one another. It is not my intention to argue, for example, about the process of cognition and the possibility of formalizing it on the basis of algorithms (something done in Appendix 4–1), nor about the models which interpret how sensitivity towards the use of norms increases the rate of genetic assimilation (p. 290 ff.), nor about any of the similar matters which inundate the book with mathematical formulae. The truth is I cannot face it, since what seems to me to be dubious in the book is not the fact of formalization in itself, but rather of certain prior questions which refer to the model of interrelation between genes and culture which it proposes. Supporting me in this is the fact that Wilson himself affirms that reduction is the key weapon of scientists.

The proposal of biological determinism in GMC is very different from the one which figures in Wilson's previous books. An attempt is made to resolve all the vagueness about which are the determinist mechanisms through a translucent box model of how genes act on the behaviour of individuals of different cultural species. The body of theses constructed to do this is called 'Comparative Social Theory' by the authors (CST from here onwards),[7] and supposes according to them an alternative to other theories which accept a cultural autonomy from genetic evolution (theories which maintain that evolution is either limited to creating the conditions and capacities for culture to appear and develop on its own, or is a process with a time and rhythm of behaviour completely different from that of cultural change). CST claims that human culture is nothing more than a further case within the series of cultural species. The features of genetic-cultural properties can be deduced therefore from a general matrix which goes beyond the arbitrary limits of human variation (p. 2). However, human culture does have at least one differentiating feature which derives from our being the only species which is at a sufficiently elevated state of the five proposed by the authors, the 'eucultural' (pp. 3 ff.). Needless to say perhaps, that does not in principle confer any privilege of autonomy.[8]

The most significant element which characterizes culture in CST is the *culturgene*, which is defined as the series of behaviours, artifacts, and *mentifacts* (in Huxley's sense), all of which are transmissible (p. 27). What is specific and original in CST is the form in which culturgenes are transmitted. This is not done in any old manner, but through a sequence of 'epigenetic rules', rules which are themselves genetically controlled. Rates of individual change from one activity to another, as a consequence of the culture which surrounds the individual, are determined in human beings by a net of epigenetic rules which operate on the data gained through learning and cognition. As a result of these epigenetic rules, innate functions appear which are in turn responsible for generating the social norms of conduct through which reproductive success is achieved. Norms evolve in the classic manner, that is, in response to special pressures of selection, but the key to such a proposition is to be found in the specific way of postulating the relation between rates of cultural change and epigenetic rules. Thus the possibilities of the CST model lie in fact in the significance of epigenetic rules and the way they can determine genetically the channel of cultural development by means of culturgenes and their mobility. This is something which GMC expresses through the use of mathematical formulae about the probability expressed through bias curves (pp. 7 ff.). Later we shall talk about such curves. First let us continue with epigenetic rules and their role in the control of human culture. GMC poses the possibility of 'genes–culture co-evolution' in the human species as the crucial question, and examines the conditions necessary for this to exist. Needless to say, such conditions are precisely those contained in the CST model, so I shall not pause over them. I should like to point out the principal empirical problem which is raised once the theoretical conditions are resolved: what degree of rigidity and specificity of epigenetic rules has been created by genetic and cultural co-evolution? – since bias curves, as we shall see later, can admit very different types of genetic programming and can even ignore it. It is this examination of epigenetic rules which can in fact solve the real problem.

Epigenetic rules are the channels of expression of genes, within the gene-environment process of interaction; channels which affect conduct through complex sequences of events occuring in the nervous system. Put that way, it seems that we are not really clarifying what they consist of, but GMC proposes moreover a far more explicit classification (p. 36):

1. Primary epigenetic rules, consisting of those more automatic processes able to lead from sensory filters to the perception of information coming from the environment. So, for example, we perceive four basic colours.
2. Secondary epigenetic rules, acting on information from the

fields of perception. They include things like the evaluation of
what one perceives and decision-making, where individuals
are disposed to use some culturgenes rather than others. A
secondary epigenetic rule would, for the authors of GMC,
explain why children aged between 6 and 18 months show a
phobia of strangers.

Primary epigenetic rules would be particularly interesting in
clarifying our question of what is the significance of genetic
determination, not only at the gamma-moral level, but at all the
others, providing we could establish the relationship maintained
with the structural elements of language. Unfortunately, GMC
claims not to have studied the relation between deep grammar
and epigenetic rules yet, and limits itself to stating that, in the
short term, epigenetic rules appear to affect the evolution of
language, while, in the very long term, cultural pressure towards
the expansion of human language manages to model genetic
evolution itself, at least as far as epigenetic rules themselves are
concerned. But this does not say very much. That deep grammar
should depend on genetic inheritance is evident within the scope
of Chomskyan hypotheses; that the existence of expanding linguistic
capacity must have carried a decisive weight in the very process of
genetic evolution and adaptation to the environment is a conclusion
as obvious as it is well known. Thus, primary epigenetic rules hardly
serve to transcend the explanatory level of the dark box models like
Chomsky's or its derivatives in Lenneberg. What about the secondary rules?.
    Secondary epigenetic rules do contain appreciable novelty. At
the outset they are enormously significant for decision-making in
the mind (pp. 51 ff.). According to CST, the mind does not act on
all information contained in culturgenes. Decisions are based on
certain features extracted from the final representations of culturgenes in the perceptive space, which is the same as saying that
the brain makes 'simplified models' of information received and
does so according to certain norms supplied by culturgenes. Only
in this way are the characteristics of speed, precision, and sim-
plicity guaranteed, which are detected by ethologists and prac-
titioners of cognitive psychology in the way individuals understand
information coming from the surrounding environment. GMC

adds the explanation of how such simplified models are made – through a process in which culturgenes play a role of the utmost importance. Thus a formulation of the determinist content of CST is in fact reached. The habitat/culture relation, expressed in culturgenes, is presented through a model contrasted with that of historical and cultural materialism (p. 56). For the authors of GMC both materialisms suppose that it is the infrastructure which determines primary culturgenes, from which the others are derived. I am not going to go into this, although I must express my deep disagreement with such a reductionist conclusion, since the important thing is to evaluate the alternative model of CST itself. This means realising that neither economic strategy nor habitat are the principle motors of human behaviour. They represent only limited conditions whose selection is influenced by epigenetic rules. These restrict rather than direct the possible choice of individuals, but their importance in this respect is enormous because it means that they manage to mould the general trajectories of cultural evolution. Let us see how.

We had established at the beginning the existence of bias curves which represent the probability of alternative use of various culturgenes. CST admits the existence of three types of bias curve (p. 9):

- the purely genetic. A property of tightly programmed species: only a particular culturgene can be chosen.
- the purely cultural. Completely foreign to programming. Any culturgene can be chosen without any influence whatsoever from the genetic structures.
- the gene-cultural transmission curve. With multiple possibilities of choice, but the culturgenes do not have equal chances of being chosen.

These curves are the product of epigenetic rules which each individual of the species possesses. The third, which is really significant in CST, is the one considered to be present in all species with cultural behaviour of any degree. In fact, the limited curves 1 and 2 only represent lines towards which gene-cultural curves can tend, which in fact can be extremely rigid (like those derived from primary epigenetic rules), or very variable. In any

case, these curves mean that the possibility of learning in certain
culturgenes, the preference of some over others after perhaps
multiple reflections, and the very use of culturgenes are phenomena which all depend on epigenetic rules. Even the most flexible
curve, capable of allowing the individual the same possibilities of
choosing almost all the culturgenes, would be a product of genetic
programming expressed by an epigenetic rule. But I believe that
such an extreme supposition poses a serious problem for CST.
The variation of cultures, according to Lumsden and Wilson,
does not of itself means the absence of epigenetic structure;
genetic control determines that the probability of use φ between
two culturgenes can be expressed thus:

$$\varphi = U_2 - U_1 \qquad \varphi \text{ varying between} \qquad -1 \leq \varphi \leq +1$$

and one of two things can happen: the probability of use φ may
remain constant when both the environment and the culture
changes, or φ varies through different norms as the context alters.
In the second case, and providing we consider sufficiently flexible
bias curves, we find ourselves in a perspective very similar to that
which CST dismisses as mistaken and tries to substitute: that of
genetic programming limited to producing the necessary conditions for the appearance of culture ('theory of the Promethean
gene' for Lumsden and Wilson). Cultural materialism itself would
in fact be a similar model. The problem is aggravated for CST
because to offer a sufficiently plausible model it has to accept the
existence of a series of principles (the *leash principle*, the *principle of parsimony*, and the *transparency principle*) which appreciably limit the rigidity of determinism.

The leash principle (p. 13) is, metaphorically speaking, the
symbolic tether which controls the tendency to use certain culturgenes capable of leading to key features contributing to genetic fitness. So, CST admits that in eucultural species like our
own, the leash-effect is very flexible due to factors linked to the
metabolic cost. For its part, the principle of parsimony (p. 69)
establishes that human beings have followed a moderate path
phylogenetically in the evolution of genetic rules: evolution is
delayed when the rules reach the minimum degree of selectivity
required. This delay means that there will not be a great quantity

of special coded mechanisms for the development of social conduct, although Lumsden and Wilson maintain that it is highly
improbable that they would all be eliminated. Finally, the transparency principle (p. 70) maintains that the more the effect of a
category of behaviour on genetic fitness depends on environmental circumstances, the more clearly the conscious mind perceives
this relation, and in consequence the more flexible is the response. In an extreme case, behaviour is modified to suit each
particular contingency that comes up through a conscious reflection on the circumstances which appear. Economic conduct for
CST would be an example of extremely flexible conduct, and at
the same time of a very high selective value.

What the whole perspective of those moderating principles of
genetic determination implies is that epigenetic rules have a
degree of influence which could be too low even to sustain the
innate content which we have postulated for the alpha-moral
level, let alone the question of beta and gamma-level determinants. If we take moral values to be highly significant in the task of
adaptation, (and this is the key supposition in the theory of moral
functionalism to sustain the presence of genetic determinants),
such values would, by the transparency principle, be the most
flexible. However CST presents certain moral phenomena, such
as the taboo of incest, illustrating exactly the image of an inflex-
ible feature. Does that mean that moral values are in fact barely
significant for genetic fitness? In any case, could not cultural
tradition alone explain this degree of rigidity?

CST wavers between two distinct extremes: either it has high
explanatory power and a model of great formal content, difficult
to apply to the human species, or conversely it manages to adjust
the model to our conditions at the expense of losing those original
features which in fact represented an alternative to the most
classical supposition of the Promethean gene, together with a
problem of methodology which I must mention.

I referred earlier to the proposition contained in GMC which
considers human culture as one more within a series of all
possible cultures, expressed in a matrix. Lumsden and Wilson fall
back on the success of the natural sciences in following a method
which considers all possibilities in order to define and study some
given feature. In consequence they postulate that all cultural

features (all culturgenes, to use the terminology of GMC) significant for the human species, and the features of genetic-cultural
properties would be better studied if considered as forming part
of the maximum imaginable collection, beyond the limits of
human variation.

There are two implications here which transform CST into a
formal structuralism similar to that which Parsons in his day tried
to establish around his theory of social action: that a matrix
capable of containing all possible cultural features, human and
extra-human, is imaginable, and that the viability of human
variation in this cultural aspect has certain limits. From what we
know of the way cultural responses to environmental stimuli are
produced, there does not seem to be much basis for either of
these two suppositions. The idea of a general matrix, definitively
fixed, whose form could not be affected by any future culturgene
whose scope we cannot now even imagine, is, quite simply,
touchingly naive. To maintain the existence of limits on the
variation of human cultural solutions (beyond the limits set by the
physical condition of man, which are of a very different nature) is
simply incompatible with the quite easily sustained conception of
the creative character of human language, which Noam Chomsky,
among others, has referred to as the basis of his innatist theses. In
fact, everything that has been written against social structuralism
– the functionalist work of Parsons as well as the marxist work of
Althusser – could be brought in at this point, but I suppose we
can be spared the effort.

From now onwards I shall use the perspective offered by the
Chomskyan model of linguistic competence as a centre from
which to analyse possible biological determinants in the beta-
moral level; but not directly. I shall make use of the correlation
made by Lenneberg in *Biological Foundations of Language* (1967),
explicitly accepted by Chomsky himself in *Rules and Representations* (1980). The biological theory of the development of language stated by Lenneberg contains some essential points which
the author summarizes in Chapter 9 of his work; I shall include
here some of those which are of special significance for the
arguments to be dealt with later on.

1. Language is a consequence of certain biological peculiarities of

the human species which make a specific type of cognition possible. The cognitive function which underlies language represents the most basic, phylogenetically determined, process.

2. The biological properties of the human form of cognition establish certain limits within which different natural languages can vary, through certain features common to them all.

3. The dependency relationship of natural languages on the human form of cognition does not extend to the external forms such languages take. External forms can vary with relatively great freedom (and within the phonetic range of possibilities of the species of course).

4. The disposition towards language which human biological properties of cognition provide has the sense of a 'latent structure of language'. During ontogenesis each individual must develop his own language through 'actualization' of the latent structure; this process of language development takes place in a linguistic environment which functions as a releasing mechanism. Without actualization, the individual not only does not learn to speak, but he does not even reach standard cerebral maturity.

5. The propagation and maintenance of linguistic conduct, although taking place within social groups, is not comparable to the cultural tradition which is transmitted from one generation to another. The individual, from the linguistic point of view, constitutes an autonomous unit which constructs language for itself, although under twofold influence from the group: (a) it represents the 'liberating environment' of the process of language and (b) it models the external form of language which is going to develop, and some elements of the structure of conduct, through the superficial adaptation of each individual.

All these points lend a base from which to interpret biological determinants, which can appear at different moral levels through linguistic conduct. There remain however some particularly dubious points to clear up: the weight which the biological properties of the form of cognition carry in the structure of beta-moral criterion and the significance of that 'relatively great' freedom which empirical gamma-moral norms can attain, a freedom threatened by not exclusively linguistic determinants from the perspective of sociobiology.[9]

CHAPTER 4

THE BETA-MORAL LEVEL. THE GOOD AND
THE YELLOW

> Now the question of how much our ordinary schemes depend on innate brain structure is what I call the Kantian problem, and it seems to me that it will be solved eventually by neurophysiology and psycholinguistics. Of course even if it should be shown that our ordinary conceptual scheme of substance and change is innate to us, this does not mean that it is necessarily a good one. It means that it has proved useful in the history of the race, not that it is metaphysically adequate. We noted in Chapter 3 that an incorrect theory can sometimes be more useful, practically, than a correct one.

> J. J. C. Smart, *Between Science and Philosophy* (1968)

The closing words to Smart's famous book on the philosophy of science gave me an uncomfortable feeling when I read them for the first time. Wittgenstein in his day had proclaimed it impossible to construct ethics in a strict sense precisely because of lack of evidence about the psychological structure of man, but, according to Smart, even Wittgenstein's caution actually turned out to be the height of optimism. Even if one day such evidence about our ordinary conceptual schemes should exist, it seems that we would have to content ourselves with having solved the mysteries of our ontogenetic and phylogenetic history. The moral matter would still be dependent on what is metaphysically adequate.

But, what on earth is metaphysically adequate? How can we recognize that something, whatever it might be, has that characteristic? Perhaps Smart could give us a lead: he refers, if we heed him, to something beyond our ordinary schemes dependent on the structure of the brain. But, as far as being optimistic goes, those who derive optimism from such criteria take the biscuit. How can we be so sure that moral language falls outside such structures? Or are we claiming that something necessarily good exists and as such is opposed to that which is contingently good, which would indeed be linked to cerebral matter? But, if the latter is so, have we not turned the Kantian problem into an absurd piece of nonsense? Because then, this necessary ideal of goodness could not even be formulated in our nearest and contingent terms. Or perhaps it could as long as we were willing to admit the collaboration of a *Deus ex machina* capable of getting us out of the difficulty. In that case he would also need to have sufficient powers of persuasion to explain that the relation existing between science and philosophy runs along these lines.

I imagine that Smart was in no way trying to pose matters in this way. To my way of thinking what has happened is that from indications which are sufficient to define the terms of discussion on biological determinants of moral language he refuses to draw the ultimate conclusion and make a declaration of ethical naturalism. That 'yes, but no' ends up by treading dangerously ambiguous ground, but undoubtedly there are those who prefer to be accused of ambiguity than fallacy.

In this chapter I am not going to dodge either of the two accusations, but if I must choose I would rather be guilty of the naturalist sin, something which I will moreover try to defend. What I support is the (partial) connection existing between the biological substrata on which language and the content of the beta-moral level depend, expressed above all by the evaluative term 'good'. If such a thesis proves to be convincing, Smart's paragraph could be rewritten, this time identifying what is useful in phylogenetic terms of moral language with what is metaphysically adequate, apart from certain doses of independence which will appear later. The strategy used to achieve this will be to consider evidence from biology as well as the logic of moral discourse itself.

Perhaps now would be a good moment to say a few words about the naturalistic fallacy. It is quite a well known fact that moral philosophers, or at least those who, as well as belonging to that category of thinkers, come under the umbrella of analytic philosophy, have for almost eighty years been pointing out to the rest of mortals interested in the question the subtle shades of meaning which link and separate the adjectives 'good' and 'yellow'. I am not threatening to go on repeating here the at times conflicting arguments which support G. E. Moore's discovery and perhaps anticipate the British moralists of the 18th Century, or Henry Sidgwick himself. But I will allow myself to recall, not without blushing for the triteness of the theme, that since that time, anyone who defines terms with moral sense (such as 'good' or 'just') by reference to natural characteristics which may include such different things as happiness, divine will, or the well-being of the philosopher determined to be mistaken, is accused of committing the naturalistic fallacy. This admirable consensus is interrupted if we try to pass terms such as 'yellow' through the same sieve, but I trust I may be allowed to affect ignorance of the outcome of this debate, which is somewhat secondary to the prime objective of discovering what happens to the language related to 'goodness'.

I am not very sure that the ethical theory which I am going to defend can be defined as *naturalist*, at least in the sense of the logical fallacy. The truth is that since the appearance of Hare's prescriptivism and the discovery of the supervenient character which language enjoys, it is ever more difficult to be fallacious in the classic sense. What is beyond doubt is the need to understand the significance of innate (that is, natural) determinants in evaluative terms of the human language. It seems that to distinguish between (natural) properties and (non-natural) values is not a particularly useful route,[1] given of course that the field of knowledge is sufficiently clear on what relates to the former. 'Good' and 'yellow' are epithets applied to things around us. This platitude is merely trying to establish that the entire scientific task directed at knowing what surrounds us has suffered such defects in the last decades that there does not seem to be any privilege of certainty and immobility attached to genuinely natural properties. It is the matter of knowledge *en bloc* which can be called into question.

But there is a sense in which (if we eliminate metaphorical use) 'good' is unequivocally different from 'yellow': it is that which includes the condition of referring to a human, or at least an

anthropomorphical, action. Even current attempts to transfer ethical values to the animal kingdom with the appearance of 'animal
rights' only refer to the existing relationship between man and
other animal species. As far as I know, nobody speaks of the
moral duty a group of lions might have towards the zebras they
try to hunt, and to bring in the 'ethics of termites', as I mentioned
in the first chapter, is the result of a misleading use of analogy. It
is probably Ayer who has most concisely expressed the significance of this supposed presence when he states that what can be
described as moral attitudes consists of certain norms of conduct,
and that the expression of a moral judgement is an element of this
norm.[2] Norms of conduct obviously lack meaning without a
reference environment, which necessarily includes at least one
human group.

The next chapters will discuss the manner in which, in the
opinion of sociobiologists, the presence of the group can imply
the appearance of natural determinants at the gamma-moral
level, the level of empirical moral codes, and the way a competitive kind of morals (such as in ancient Greek times) identifies
values and natural properties closely linked to the situation of
social class. Given that little is gained by merely stating that
aristocratic Greek morals are fallaciously naturalistic, those who
persist in maintaining the analytic point of view usually invoke a
question of principle: the theme of the irreducibility between
values and properties belongs to meta-ethics, the analysis of
ancient morals does not. Of course I, too, am going to defend the
separation between the beta-moral level (which considers such
relations) and the gamma-moral level (about the positive contents of a given ethical system like the ancient one). In no way do
I claim to establish that those levels form watertight compart-
ments, impossible to link. On the contrary, the position I take
claims to demonstrate that the paragraph by Smart at the beginning of this chapter is erroneous in insisting on the formal separation. My doubts about the technique of escaping to the plane of
meta-ethics, if what is under discussion is the meaning, significance, and possibilities of the phenomenon of beta-moral evaluation, can be illustrated through what is implied by the mutual
influence between reason and action.

Toulmin (1960, p. 152) points out that there are four ways of approaching the problem: historically, psychologically, logically or philosophically. Even so, for him "Answers to all these forms of the question [the first two, C.J.C.C.] will be interesting as illustrating what we are after, but the nature of logical criteria applicable to ethics is manifestly independent of them". If what Toulmin is saying is that the logical approach to ethics only coincides directly with the logical approach, there is little to discuss. But that is not the aim, rather it is "how is it that we come to let reasoning affect what we do at all?" The approach termed logical supposedly lies in "an enquiry about the kinds of changes in behaviour characteristic of a decision based on 'moral' grounds; about the way in which reasoning must be designed to influence behaviour if it is to be called 'ethical'; and so on." However, precisely because that is the line of investigation considered adequate (given that the essential problem is the way reason affects conduct), it does not seem that the whole matter can be restricted to the terms of the logic of statements by agreeing that this, in fact, is the place for moral foundations. Rather we would have to explain how Socrates allows reason to affect what he does, under the inevitable imperative that it is Socrates himself who claims to be dealing with moral matters. To claim, with Toulmin, that the truth of what we are now after must be independent even of whether Socrates existed remains an academic conjuring trick not too far removed from Smart's claims for adequate metaphysics.

Perhaps norms of conduct have a universal component (or perhaps not, it is up to biology and ethology to tell us); undoubtedly there is a logic to moral statements. But in any case the way in which Socratic reasoning affects norms of conducts in the 4th Century *polis*, or for that matter, in Enlightened European society does not seem by definition to be absolutely beyond any type of biological determinant. I am not claiming that gamma-moral contents rely on such determinants, but that beta-moral discourse is what remains linked, in a way which remains to be shown.

To clarify how that type of determinism could appear in moral discourse, we shall begin by posing a question which at first sight does not appear to be related to what we are after. We are going to try once more to investigate what the use of an evaluative term such as 'good' means. The philosophy of moral language has been finding a series of characteristics in this type of evaluative term which are derived, in general, from the idea held by each school about the essence of the term 'good'. Thus, it has been claimed that 'good' is a simple notion (Moore), or in any case ultimately reducible to an intuition (Prichard), or, if preferred, to a tendency of the mind (Wittgenstein); also, that 'good' expresses a certain feeling or attitude (Ayer and Stevenson), or that it has a sense of recommendation or guide to action (Hare); or finally that it refers to the content of the evaluation and not to the form it takes (Foot), with which, if we go along a little with this, we return to the Aristotelian theory of virtue. None of these stances has managed to come through the barrier of their respective critics unhurt, unless we resort to the oft-used solution of the conflict of paradigms, establishing that all those authors share a sort of internal truth, which is not far from saying that they are all wrong. If a general equivocation has to be conceded, then we will have got into serious difficulties because it would be difficult to reason in greater detail than the analysis of moral language has already done. To accept that the correct way of analysing the richness of evaluative language has not been found could be somewhat closer to having to admit that terms such as 'good' naturally contain something indefinite *more analitico*. I am far from proposing a synthetic solution, but it seems to me that the idea of insuperable ignorance is excessively strong and that the sense of evaluation could indeed be understood as long as we agree to avoid the reductionist temptation and a flight to the field of the formal, accepting in consequence that evaluative words such as 'good' have a multiple nature.

To make progress I think that as a first step we can leave aside emotive elements present in moral language, not because they are not used in such a way (something which undoubtedly happens) but because of the need to understand that emotive use distorts the presence of evaluations, or rather, transforms the basic intention, moving it into the sphere of rhetoric. Given that these days rhetoric once more demands its place beneath the skies of ethical discourse, and is another of those facets taken for dead and buried which have regained their old vigour, all we will do here is provisionally leave the matter aside to return to it later, for *what we will now discuss is the possibility of putting forward rationally founded criteria of ethical preference*. The 'magnetism' of moral terms is to a large extent concerned with the empirical question of how we manage to convince our neighbours of the necessity of acting in a certain way, and that is something which will have to be taken into account. But the truth is that such magnetism loses a lot of its force when what is proposed is a discussion about the possible validity of a moral judgement made by a third person. On the other hand I do not wish to give the impression that, like Stevenson, I think the ambiguity of terms such as 'good' grants them a certain immunity from rationally cognoscible explanations about their character. If something like this has happened, it is due to my own ambiguity: when I spoke of the multiple character of evaluative terms I meant the need to consider them as fulfilling different functions in their everyday use, and the possible confusion between all these functions leads us to necessarily ambiguous standpoints from the moment we claim that words such as 'good' only serve a single purpose.

In spite of all that has been written regarding a taxonomy of linguistic uses, the classification made by Popper in his speech given in memory of Arthur Holly Compton in 1965 still seems to me to be of a profound clarity; a classification which in part expounds some theses already formulated by Karl Buhler as early as 1919. Language (in general) is a form of behaviour whose functional character is shown in four distinct ways through:

1. The symptomatic or expressive function, which reveals the state of the organism from which linguistic signs proceed.
2. The releasing function. Through the previous expressive function a certain reaction is stimulated or triggered in a second organism which receives the linguistic signs. (This function, and the previous one, would for Popper be 'inferior', in the sense that animal communication also relies on them, as opposed to the 'superior' character of the following two functions which are exclusive to the human language).
3. The descriptive function, whose statements, which I imagine there is no need to clarify, can be true or false.
4. The argumentative function, connected to a critical and

rational attitude which can cast doubt on the propositions coming from the descriptive function (Popper, 1965, paragraph XIV).

Popper considers that "the argumentative function of language has created what is perhaps the most powerful tool for biological adaptation which has ever emerged in the course of organic evolution". This is perhaps a rather anthropocentric statement because doubtless there are extremely powerful instruments of adaptation, such as the presence of a high-level genetic variability, which are patently able to compete with critical language, if we take but little account of what happens in influenza epidemics. I suppose that the ranks of ethologists would also expound reasoned doubts about the classification between superior and inferior functions, and the absence of the descriptive function in animal communication. But that can all be left aside since what is really relevant to our purpose is to observe the evidence recalled by Popper that critical language is one of the main instruments in the biological adaptation of our species, and to call attention to the value of the second of the above functions (the releasing function) during the phylogenetic process, at the point both the species and the language/thought mechanism are themselves on the evolutionary path towards their current state.

The thesis I want to propound is that if we accept that the alpha-moral level is biologically determined, (and that seems certain) then so too is the beta-moral level, at least partially, through the releasing function of evaluative words such as 'good' and through the functional character of moral elements.

Ethology has developed the concept of the 'innate releasing mechanism' to explain the capacity of beings to react to certain key stimuli coming from the surrounding environment, in such a way that the behaviourist explanations like the stimulus/response model cannot be applied to these reactions. In this case experience is not relevant, at least initially, to the reaction. An example from Eibl-Eibesfeldt can serve as an illustration: when a tadpole which has survived by feeding on algae undergoes its metamorphosis and becomes a frog, it tries with a flick of its tongue to trap everything which flies. In this way it gets leaves, small stones and of course insects. Later on it learns to discriminate and only goes after what is really edible. But at first it is the presence of moving objects which, as a key stimulus, releases conduct which we consider of necessity to be innate. In this sense Eibl-Eibesfeldt considers that innate knowledge exists (1967, pp. 94–95).

Are there releasing mechanisms of this type in the human species? This is a question which would have to be linked to the wider discussion of whether there is any type of instinct in the behaviour of human beings, which has typically produced two different replies. Cultural anthropologists, led by Ashley Montagu, have put their efforts into demonstrating how hominization would be a process precisely where the functional character of instincts is lost. Tinbergen and Lorenz are the founders of the ethological school which sees no reason to make exceptions for human beings regarding instinctive behaviour. Logically I am not going to come out with a collection of quotes to reinforce one side of the argument or the other; it is enough to recall the change of attitude culturalists have had to adopt as ethology has begun to overwhelm the audience interested in the controversy with masses of experimental data and ever-more polished theories.[3] Today on the other hand, the nature/nurture dichotomy has lost most of its meaning, which had given rise to innumerable discussions since the now distant time when things that existed were counterposed by *physis* or *nomos*, by nature or by custom. Sociobiology and ethology have at least managed to give a different appearance to the comparison between the innate and the learned, since the very capacity to learn is considered one of the most significant elements in innatism.[4] But the source of speculation has not diminished with the apparent victory of innatist theses, and it is precisely the stance of Eibl-Eibesfeldt, linking releasing mechanisms to the existence of innate knowledge, which I should like to assess. If there is such a thing as knowledge prior to experience, how far does it reach in the human species?

Forming part of the innate stock of knowledge are such different things as facial expressions after tasting sweet, acid, and bitter flavours; reactions to objects moving towards the observer; the perception of spatial and temporal evidence; the subject's attitude to the space around him; a friendly disposition to clearly child-like features (large head, convex forehead, big eyes); understanding of facial expressions (pride, rejection, friendship) and others along these lines. But if we pay heed to Eibl-Eibesfeldt, who in turn follows Konrad Lorenz on this point, certain situations in human co-existence would act as releasing stimuli capable of causing our ethical value judgement to react; this implies the existence of a certain pre-programmed, or if preferred, simply a programmed ethical conduct.[5] This thesis could be interpreted in a weak way, by referring only to ethical programming contained in the alpha-moral level, which would explain the obvious fact that we are beings with ethical behaviour. But I should like to mention here a more forceful reading of ethical pre-programming: that which accepts the presence of releasing mechanisms at the beta-moral level, directly linked to the world of moral discourse.

The phylogenetic process of incorporating linguistic communication, at least in its early stages, could be understood apart from the functional nature of language with regard to adaptation. One could think that the capacity to communicate at an advanced level appeared independently of the mechanism of natural selection, through genetic drift and in the form of a totally random, stochastic process, linked to neutral genes.[6] If one admits that the first moments of hominization occurred in a very limited, sparsely populated environment, that exceptional character could have existed however much the relatively high level of communication among superior apes might make such a hypothesis quite implausible. Even so, the development of incipient capacities can in no way be detached from the immense functional value of a sophisticated means of transmitting information. Even if the speed of genetic incorporation, the rhythm of phylogenetic acquisition, could have varied, there are certainly three types of capacity which would have to have been co-ordinated at a given moment to result in the human language: the capacity to emit discrete phonetic signals, the capacity to order the world through very precise semantic contents, and the capacity to produce phonetic/semantic connections. A large part of the final result, in addition to the capacities themselves, seems innate. There are obvious exceptions, like the peculiar type of phonetic/semantic link whose empirical expression forms the content of the different languages which exist, but the structure of the phonetic/semantic link itself has been considered since Chomsky to be connected to innate tendencies.

Let us imagine, then, the situation of some hominids who are developing their linguistic capacity and already have, moreover, a sufficiently high level for language to figure as a functional element of adaptation. From Lieberman's research (1973) this supposition at least can be attributed to Steinheim's fossils. It is to be expected that the possibility of transmitting knowledge would not reach very high levels, but what would it mean if, in these conditions, an individual from the group were able to communicate to another information about what is tangible or rejectable in a moral sense, however primitive that might be?

This question is an attempt to understand the role the concept 'good' might play as a releasing impulse.[7] Two extreme positions are possible: either the linguistic use in this sense only relates to argumentative behaviour of a rational type, or, in contrast, it can also entail the possibility of releasing instinctive response mechanisms. In either of the two cases what is beyond all doubt is the high functional value of the adaptive character of moral language, linked precisely to primary concepts, that is, those of little sophistication. It is this characteristic adaptive value which casts the greatest suspicion of innateness on language. Konrad Lorenz for example, maintains that the functional character of moral language can only be understood if it is linked to automatic responses, because the situations in which it would have to demonstrate its effectiveness are just those where there is no time for discussion about why certain things should be done or avoided. By a cultural route, primitive communication had developed weapons which greatly transcended the capacity of, shall we say, natural weapons – teeth and bare hands – to injure and kill, so that one of the fundamental principles of ethology (that of the proportional relationship between the damage that can be caused and the inhibition of aggressive behaviour, which gives species such as predators, who are able to kill with relative ease, strict mechanisms of containing aggression) would have been transgressed, thereby entailing serious difficulty for the survival of hominids (1963, p. 267). It is not a case of circular argument, because nothing forces us to assume that lost balance *necessarily* had to be restored. What Lorenz claims to explain is how, in fact, the species of hominids survived since the oldowan cultural traditions, when they could rapidly have become extinct. One possible response from the culturalist perspective would be to deny the existence of strong aggressive impulses in *Homo habilis*, but the truth is that there is such disproportion between cultural and biological weapons among the *australopithecus* (and of course among all later hominids) that one would have to imagine a degree of aggressive behaviour far smaller than that found empirically today in all known human groups. It was Eibl-Eibesfeldt who showed the error underlying the myth of 'non-aggressive societies' and placed the question of innate aggression in the human being in a perspective which is more in accordance with the original theses of ethology.[8]

In reply to the question of how primitive hominids could survive, Lorenz presents the theory of the biological imperative as an alternative to Kant's Categorical Imperative (1963, p. 275 ff.): man's moral conduct can only have adaptive significance if it is capable of provoking automatic responses, and these are only sufficiently assured through instinctive behaviour, a biological imperative able to contain the principles necessary to restore the balance between aggressiveness, lethal weapons, and lack of inhibitions. I think Lorenz does not do enough justice to the Kantian Categorical Imperative, but although his alternative might not seem too plausible today, the need still remains to understand what role ethical behaviour plays in the phylogenetic process, a role which has been claimed at least since Darwin's initial formulation of the essential norms of biological moral functionalism.

We are trying to establish the way in which alpha-moral level behaviour forms its link with the beta-moral level, and to what extent this is done through the presence of innate elements, of biological determinants. If moral conduct is genetically defined as that capacity, is there also a system of meaning, a meaning to moral language which could also be considered innate? An affirmative response would oblige us to postulate the presence of universal structures in language capable of directing alpha-moral conduct in some given sense. But that is equivalent in practice to admitting that the terms of moral language have, at least partially, a releasing function. And the question is somewhat subtle, since we are dealing with a very complex releasing mechanism. Releasing mechanisms detected by ethologists in many cases imply voiced signals, but the response is codified by virtue of standard phonetic signs for each species. This is not the case with moral language. There is nothing similar to universal phonetic signs which could be interpreted as a means of inhibiting aggressiveness. If they existed there would be no need at all to resort to the beta-moral level; alpha conduct itself would be enough. And, in passing, neither would there seem to be any need to postulate a theory about the biological imperative, because there would be sufficient means of inhibiting aggressive conduct in the classic way. One ritual of inhibition would simply have been substituted, like any of the courtship dances, for another of a phonetic type. What appears now is a *semantic* content which acquires the character of a releasing mechanism. There would of course be other non-linguistic releasers, such as facial expressions, gestures of greeting, infantile features and so forth, but the functionalist argument gives moral language a primary role in the possibility of forming cohesive groups. And thus, as I stated earlier, if we accept that the alpha-moral level is biologically determined, we find it necessary to conclude that so too is the beta-moral level, through the triggering function of certain words with moral content such as 'good'.

The innate content of the evaluative word is not easy to define. It is obvious that no specific phonetic/semantic combination can have this characteristic. Neither will the concrete use, the application of the quality of goodness to some given acts have it. An English child learns what is good and what is not through endoculturation within a group whose social and historical characteristics lead him eventually to speak of 'good' things and 'goodness' (if he is an English speaker, of course) and not of 'bondad' as if he were Spanish, and to apply such terms to acts which would perhaps be considered differently by a different generation, ideological group, or social class. To assume that there is a specific content to 'good' applicable to any concrete value is to extend genetic determination to the gamma-moral level, and we will discuss the viability of such a thesis later. The only thing we are trying to verify is in what sense is there determinism in ordering the world around a dichotomy like good/bad.

One possible lead about the meaning of beta-moral determinism may appear if we use the concept of 'primordial knowledge' coined by Avrum Stroll (1982). Primordial knowledge forms part of each subject's way of establishing himself in his own world. It refers to things we hold to be true, but not in the same way as we consider scientific knowledge to be true. We know that the Earth has existed for many years, that many different types of things exist in the world and that many human beings have feelings, but Stroll insists that this type of knowledge is different from that which assures us that the Earth has existed for 4,000 million years, that all things in the world are formed by elements found in the Periodic Table, or that the working of the human brain takes the form of a feed-back mechanism in some of its functions, a mechanism similar to that of digital computers (1982, p. 181). Primordial knowledge must fulfil certain conditions (as it has to be absorbed and not learned its denial implies an absurd description of the world and is an 'absolute language game' in the sense that the propositions which such knowledge contains are fundamental or basic) which distinguish it from scientific knowledge. Stroll insists on the fact that primordial knowledge is of an empirical, not analytical, nature: we know the world primordially because we are immersed in it. Such a characteristic could lead one to think that this has little to do with the innate moral knowledge we are trying to discover. Once again it would be an error of approach to agree with this. The parallel I am trying to establish between 'primordial knowledge' and 'innate moral knowledge' aims above all to identify the outline and function of both as necessary elements to be established in the environment which habitually surrounds any human being, and which obviously includes a social group. Something is learned about the earth and things empirically, certain moral norms of conduct of a positive nature present in the group in question also become known empirically. This is something related to the gamma-moral level. But what concerns us now, beta-moral content, is also related to Stroll's primordial knowledge, since it provides the tools necessary to form a bridge between human alpha-moral nature and the concrete values of each human group. The tendency towards moral conduct is specific to the human being and impossible to compare with that of other gregarious species like termites, precisely because of the presence of the beta-moral level within certain linguistic behaviour; linguistic behaviour which contains the structures necessary for any human group to assign to the actions of its members a positive or negative moral evaluation by virtue of values which may be different, but which are present as such in all human groups.[9]

I believe that the significance of innateness in evaluative concepts lies in that structural aspect which is somewhat different from the 'biological imperative' referred to by Lorenz, for reasons which I shall now point out. Let me first go into one question. If the *whole* content of the beta-moral level were innate in origin, we would find ourselves with universal values in the strictest sense of the word claimed by natural law. A universal value of this type could be the prohibition of incest or the avoidance of murder. On this Lorenz's argument rests, and it is of course far more sophisticated than the way I now present it. If that idea were so, there would be no sense in continuing our analysis of the beta-moral level; we could go on directly to those gamma-moral elements which are universal. If I insist on the relevance of a beta-moral approach however, it is because in it will appear, together with the condition of innateness linked to evaluative concepts like that which in English is called 'good', an element of autonomy, of indeterminism. The significance of the meaning of 'good' will be dealt with later when we reach the subject of moral progress. Let us now go on with the beta-moral autonomy which we could state in the following manner:

Due to phylogenetic characteristics of human adaptation to the environment, descriptive and argumentative functions add a second aspect of indeterminism to the innate releasing nature of moral language

What this second thesis claims about the character of beta-moral elements is to make the presence of innate bases capable of justifying the existence of intuitions, such as that which leads to the primary concept of 'good', compatible with the type of behaviour which was presumably adopted in the course of the evolutionary process, behaviour which necessarily had to rely on the critical characteristics of language as an element of prime importance for adaptation. At first sight there seem to be serious difficulties in making such diverse assumptions compatible, for example:

1. The presence of innate aggressive impulses to such a high degree as to serve as a functional element in the formation of hierarchies in pre-human groups.
2. A relatively weak store of innate means designed to inhibit aggressive conduct (due to the absence of natural weapons).
3. Lethal instruments which rapidly evolve in efficiency thanks to the accumulation of cultural traditions.
4. Elements of controlling behaviour based at least partially on moral innateness.
5. Cerebral structures which lead to the adoption of non-automatic, critical responses to stimuli from the environment.
6. Linguistic structures which contain innate essentials about the way in which knowledge of the world and its evaluation is organized.

Is it possible to define an evolutionary stable strategy[10] compatible with all these basic points? To answer yes because, after all, we are discussing these things and that is proof enough that our species has managed up till now to adapt to extremely varied circumstances would be to resort to a *petitio principii*. It could be that our survival is due to something different and that some of these so-called data are contaminated from the outset. But what we know about the meaning of human phylogenesis does not seem to deviate too much from those points which we assume to be of great significance for the process of hominization.

The key to the autonomy of beta-moral elements is intimately linked to what from a phylogenetic point of view is the type of adaptive strategy adopted by the human species and is usually classified as 'open programmes of conduct'.[11] In considering open programmes of conduct we must bring up a matter which will be treated more thoroughly in the chapter devoted to the delta-moral level: that of the ultimate end of living beings. Adaptation to an environment gains biological meaning thanks to the introduction of a teleological idea capable of interpreting the 'logique du vivant'.[12] Molecular genetics identifies life with the duplication of nucleic acids, governed, in Monod's terms, by a law of invariability which converts individuals of different species into something like machines, well-designed to maintain the 'information' content of nucleic acids of genes as appreciably invariable. I shall also leave until later the possible parallel between the model of the transmission of genetic information and moral events. What I want to emphasize now is adaptation to the environment as an expression of the purpose that we assign to living beings. For the teleological project to be accomplished, it is necessary for the individuals of each species to develop a certain conduct which has been sanctioned in terms of evolution and natural selection as adequate for adaptation to the environment. Of course, not *all* individual acts have to be rigidly interpreted in this adaptive sense, but it is certain that some particularly significant acts of conduct exist in which to a great extent lie the hopes of fulfilment of the laws of teleology. I am not claiming that important acts are those which by definition have such a characteristic. I mean that the activity of living things is related to the way certain strategies are oriented to performing operations necessary for the transmission of genetic content, and that the meaning of behaviour in this sense can be hidden beneath sophisticated activities which may at first seem removed from the adaptive task. While the phenomenon of the transmission of genetic characteristics is an individual question in the final instance, what is really significant with regard to teleology is the activity of an individual in an environment where there are not only 'things' but 'individuals' of the same species, with DNA content reasonably similar to his own. But this does not necessarily mean that the behaviour of all these individuals is the same. The programme of conduct determined by genetic means may be very rigid and so give rise to extremely similar behaviour in individuals of the species in standard environmental conditions. But genetic programming of behaviour is not directly related to the rigidity of eventual conduct.[13] Genes may have led individuals to more open programmes of conduct which oblige different individuals of a species to perform different acts (above all relating to the narrow pursuit of differential circumstances in the environment), thereby introducing a new type of individual behaviour, that of choice. It must be stressed that an open programme of conduct which obliges individual choice between options of action, which are diverse and of different significance in a changing environment, is as much genetically determined as a closed programme of conduct. Until now we have in no way crossed the threshold of determinism, even though the open programme of conduct can serve as a basis for doing so. Superior mammals have much more open programmes of conduct than for example the fish who form shoals, but that is not reason enough to take the presence of moral indeterminants for granted. But the open programme of conduct in the human species has a special meaning when what is analysed from this perspective is the characteristic of language as a basic element in adaptive strategy. That implies, in more classical terms of ethical discussion, introducing the presence of will.

And here we must again tackle the discussion which began in the previous chapter with reference to the alternatives of Kantian metaphysics. Without doubt the will to act is one of the principle conditions which allow us to talk of the existence of ethical choice. Even such determinist proposals as those of sociobiology try to find a corner for human liberty and will, and reference was made previously to the interpretation of both by Edward O. Wilson. Let us recall his conception:

So, for the moment, the paradox of determinism and free will appears not only resolvable in theory, it might even be reduced in status to an empirical problem in physics and biology. We note that even if the basis of mind is truly mechanistic, it is very unlikely that any intelligence could exist with the power to predict the precise actions of a human being, as we might to a limited degree chart the path of a coin or the flight of a honeybee. The mind is too complicated a structure, and human social relations affect its decisions in too intricate and variable a manner, for the detailed histories of individual human beings to be predicted in advance by the individuals affected or by other human beings. You and I are consequently free and responsible persons in this fundamental sense. (1978, p. 77)

But this is an argument about liberty and indeterminism which Kant had already announced as a fallacy. If human freedom can

only be linked to the fact that psychic phenomena are too complex to be embraced in totality, we are completely within the
phenomenic field which corresponds to 'spurious freedom', in which
the free will necessary to lend a little weight to the ethical
framework is difficult to sustain. In fact, Wilson has not separated
himself one whit from the initial proposals of ethologists like
Lorenz and Leyhausen designed to make human will compatible
with innate releasing mechanisms. For my part, I think that
Kant's intuition about human will was much more acute than
Wilson's, that is, that the open programme of conduct as a stable
evolutionary strategy in the human species is related more to the
freedom connected to indeterminism than to that other, spurious,
freedom related to the limitations of knowledge about mechanisms of determinism. But obviously I am not going to follow
closely in Kant's footsteps and postulate ethics formally separate
from the order of causality. There are two extreme alternatives in
terms of moral conduct: to suppose, with Wilson, that the ethical
system of the human being is immensely more complex than that
of social insects, but similar in terms of its genetic subordination,
or to propose, with Kant, the existence of an unbreachable gap
between two types of entity which belong to orders impossible to
reduce between them. It seems that an intermediate path can be
defended which accepts the presence at the beta-moral level of
genetic determinants capable of channelling as a structure some
extremely important elements, but with the addition of an aspect
of indeterminism within the beta-moral level itself, linked to the
presence of critical language.

How far is this an alternative to Kantian metaphysics? Is an
autonomy capable of retaking the path towards dualism between
the orders of nature and reason perhaps not also being postulated? To answer this question it is necessary to examine in a little
more detail the meaning of the proposal contained in the two
Kantian *Critiques*.

Leslie White Beck has stressed the ambiguity of the Kantian
treatment of the subject of will and freedom. In fact, within
Kant's work two different concepts of freedom can be found. The
first, from the *Critique of Pure Reason*, allows the difference
between the natural and rational orders to be established, considering that freedom consists of the faculty to initiate new causal

series in time. Beck states that such a concept of freedom was held to be applicable to human will and spontaneous voluntary actions by Kant in 1781. But in the *Groundwork of the Metaphysic of Morals* there appears a second concept of freedom, freedom as autonomy, not as subjugation to any law except that made by man himself. To distinguish the act of will needed to spontaneously initiate causal series from that which refers to the source of the law to which such spontaneity is subject, Kant differentiates between *Willkür* (will in the sense of freedom as contained in the *Critique of Pure Reason*) and *Wille* (will related to freedom as the *Groundwork*).[14]

Willkür in fact implies the faculty to choose an object which reason shows us to be incompletely determined. If it were not so, if reason showed how determinants of such conduct exist, we would not in fact be dealing with an act of will but with conduct which is merely following the imperatives of a previous cause. In postulating *Willkür* Kant can combat any hedonism and in general any basis for that type of moral conduct, including moral sense mechanisms. If a man acts in that way impelled by psychological means, he is subject to the order of causality and his decisions can be anticipated. Neither freedom of choice nor will as *Willkür* appears here at all. But if such subjugation does not exist, if the faculty of choice remains undetermined, the object of *Willkür* may be a pragmatic action or a moral action (Beck 1965, pp. 218–219). Whether it is one or the other does not depend on the object itself, but on the rule underlying the indeterminism. If the rule of reason is referred to, which is derived from our theoretical or empirical knowledge of the causal conditions which lead to obtaining the object, (conditions which of course leave the obtaining itself indeterminate and subject to free will), then we find ourselves in the orbit of the first of the *Critiques*, confronted with a pragmatic action. But in the case where the rule of reason is directed not towards the manner of technically obtaining an object, but towards the way all actions must be universally applied to human beings, the moral law appears within the sense of the *Critique of Practical Reason*.

Willkür and *Wille* are not, then, two alternative concepts, but successive ones. Beck shows that "*Willkür* is fully spontaneous only when its action is governed by a rule given by pure practical reason, which is its legislative office" (1960, p. 180). In both cases, free will really exists and in consequence, responsibility, but only the appearance of moral autonomy represented by *Wille* and based on the fact that the author of the act of will makes the law himself allows us to speak of *moral freedom* according to Kant. *Willkür* implies another type of freedom, *comparative freedom* able to mark the difference that exists between man and animals when trying to satisfy the object of desire. As freedom which has an operative sense it can also manage to attain the status of moral, but only if the choice is subordinated to the maxims contained in the law of pure practical reason.

It seems that we are now able to appreciate the difference between Kant's metaphysics and its alternatives. I propose that we consider the following table of the relation between freedom, will, and determinism:

Freedom	*Will*	*Indeterminism*
Through ignorance of causes	sociobiological	guaranteed by the laws of physics and biology
comparative	*Willkür*	by initiating causal series
moral	*Wille*	through autonomy (identification between the author and origin of moral law)

The Kantian ethical proposal has a beta-moral sense linked to *Willkür*. But Kant did not consider this sufficient in itself, hence its ultimate meaning would not be universalizable and would remain linked to psychological explanations (causal explanations about desires). Thus he added a delta-moral sense, on which he based the autonomy of ethics, also raised as a criterion for authentic freedom and spontaneity. Kant claims in principle to reach the desired universalization, the effects of an ethical system, by means of absolute freedom. But that is not the origin of the duality of nature and reason, which had already appeared within the previous concept of *Willkür* itself and which is able to sustain comparative freedom. Autonomy is a complement to spontaneity, not an alternative. And as that is so, a norm of beta-moral indeterminism can be established without the need to accept Kantian moral freedom. Freedom linked to *Willkür* can be postulated and can combat the need for *Wille*. The final result will be a system in which we cannot resort to the Categorical Imperative as a basis for morals: universalization will have to be sustained in another type of assumption. That is why my proposal of the 'autonomy of the beta-moral element' clashes with Kantian terminology; that for Kant defines a spontaneous action, but not an autonomous one, in so far as it is subject to nature. It is not absolute subjugation, since we have established the presence of indeterminants: the subject can initiate (while in Wilson's proposal he only *thinks* he does, and nobody can contradict him). But there is an evident connection with nature in so far as it is nature, which through desires and tendencies dictates certain goals which in my proposal could be disputed. Kant considers it incongruent to claim that such a connection exists, and, at the same time, a moral obligation, but he did not have at his disposal the means to understand human nature that we have today. The 'intermediate' proposal between sociobiological will and *Wille* which I am defending attempts to use them, and in fact runs along Kantian channels close to *Willkür*. It is established a) in the presence of causal determinants linked to the genetically determined beta-moral aspect, and b) in the possibility of autonomy (in a non-Kantian sense) contained in the natural way human language functions. Understood in this way, the model of *Willkür* does not go beyond the beta-moral level itself, in contrast to that of will as *Wille* referred to as a supreme delta-moral value. And this is of the utmost importance, because as we shall see when dealing with the delta-level of moral phenomena, the proposal of ultimate values is a somewhat doubtful mechanism if one claims an absolute status for them. To remain on the beta-moral level concedes to reason the right to doubt everything when ethical judgement is being discussed. And while the Categorical Imperative does not exist to dissolve doubts, some solution will have to be found to choose between counterposed norms and criteria, a solution which will be connected to the way in which the autonomous element appears at the beta-moral level.

Despite the recovery of Cartesian linguistics made by Chomsky, ethology has managed to do away with many of the differences existing between human and animal communication. Today we know that some animal species are capable of transmitting messages of high semantic content – bees – and that our nearest relations, the chimpanzees, have sufficient neural capacity to distinguish phonetic contrasts which in human language form significant features (Lieberman, 1973, p. 94) leading scientists like Yerkes to attempt the probably impossible task of teaching some gifted chimpanzee to speak. But even supposing that the superiority of human language lies only in the quantity and sophistication of the significant features, this difference is qualitatively translated into the capacity to order a critical dialogue around the information received. The meaning of the mechanism is considerably affected by such a characteristic: the efficiency of the system's functioning can be evaluated *a posteriori*. And it is here that the possibility of autonomy lies. The critical condition of language is connected to a sentence structure (a universal and probably innate structure) in which a correspondence between the epithet 'good' (also with a universal and innate sense, as we have established up to now) and another external element, which could be a human action, is established. Perhaps that is also true of the ethics of social insects, but what is specific to human ethics is that in no way is the relationship between the term of moral content and the exterior element fixed in a rigid and definite way. Such correspondence, expounded by the individual who enunciates a moral judgement, can be cast into doubt. And that, precisely, is the origin of the gamma-moral level and the different and particular correlations between evaluative terms and elements of behaviour. At the beta-moral level, it is the flexibility of the connection which gives the key to ethical indeterminism in such a way that the existence of linguistic universals of a moral character stops at the appearance of evaluations. The possibility of casting doubt upon such evaluations and the mechanisms of discussion of ethical criteria are keys to an autonomy of discourse partly directed towards a terrain in which critical language reclaims its power.

Through the first thesis I put forward in this chapter, *'good' (as a concept) has a releasing function within the social conduct of a group*.[15] In general its use will refer to certain lines of conduct which, if we must take account of ethical functionalism, will be significant for the adaptation of the group to its environment. Describing and communicating in evaluative terms is, therefore, important with regard to the behaviour which will be adjusted in response to the information received, and this mechanism of impulse/response has been interpreted through the presence of innate structures which grant a given sense to what we express in connection with the epithet 'good'. But the second thesis, which introduces the aspect of indeterminism, changes the character of the releasing mechanism, in that now the critical element of thought and language comes fully into play around the descriptive and argumentative functions. If the urgency of the response permits (following the action, should it occur) beta-moral type lines of argument can be raised to evaluate the empirical content which under the express conditions of the group tradition come under the qualification of 'good'. That such a tradition also includes the means of blocking the discussion using ideological impositions is another matter which we will leave aside for now: what is important is that discussion relies on structural means to be produced and not that it is produced *de facto* at a given level of complexity. One must bear in mind that the express fixing of external elements in significant correspondence to the evaluative term in no way affects the structure of the beta-moral level. 'Good' is already fixed indelibly as an element of biological material and if there is anything which within a certain tradition is considered ethically desirable and then for whatever reason is doubted, the doubts do not extend to goodness itself, but to qualifying as good something which does not seem to deserve it.[16]

Recourse to the critical meaning of language as a means to support a point of beta-moral indeterminism meets however with certain derived problems. We have maintained that to rely on rational means able to criticize the content associated with the primordial concept of 'good' implies a selective advantage over automatic responses which could justify (equivocally in my opinion) the presence of ethics in non-human social species. That means that in a situation demanding moral behaviour the human being relies, at least theoretically, on rational means of deciding about what is correct or incorrect in an act to which eventually an evaluation of this type must be apportioned. The resulting critical conduct, which could in principle seem so obvious as not to warrant further comment, has been precisely one of the most problematic matters confronting moral philosophy over the last centuries. The rational character of beta-moral judgements cannot be proclaimed willy-nilly without treading the ground of rational preference, because in the not unlikely event that different subjects evaluate an action in conflicting ways, it will have to be decided whether a rational way of knowing which of the two is right, and which wrong, exists. If in the name of relativist ethics, intuitionism, or any social or biological determinism we deny the rationalization of ethical discussion, we are in fact invalidating the technique which until now has been put forward to give the beta-moral level a degree of autonomy.

CHAPTER 5

THE BETA-MORAL LEVEL: RATIONAL PREFERENCE FROM SMITH TO RAWLS

> . . . innate censors and motivators exist in the brain that deeply and unconsciously affect our ethical premises; from these roots, morality evolved as instinct. If that perception is correct, science may soon be in a position to investigate the very origin and meaning of human values, from which all ethical pronouncements and much of political practice flow.
>
> Philosophers themselves, most of whom lack an evolutionary perspective, have not devoted much time to the problem. They examine the precepts of ethical systems with reference to their consequences and not their origins. Thus John Rawls opens his influential *A Theory of Justice* (1971) with a proposition he regards as beyond dispute: 'In a just society the liberties of equal citizenship are taken as settled; the rights secured by justice are not subject to political bargaining or to the calculus of social interests.' (. . .) Like everyone else, philosophers measure their personal emotional responses to various alternatives, as though consulting a hidden oracle.

> Edward O. Wilson, *On Human Nature* (1978)

Perhaps Wilson does not do enough justice to the legions of philosophers who, following Kant's footsteps (or even before him) emphasize the deontological character of ethics and disown utilitarian calculations. I am not going to enter into the question of whether Rawls or Nozick can be rescued even partially from the virtues of a moral philosophy more attentive to the origins of ethical phenomenon than to their consequences, but I shall devote a few pages to examining the character of secret oracle which can appear behind the task of rational preference, the spirit of which is in principle far removed from any kind of magical consultation.

The individual to whom theoretically one can attribute the conditions necessary to make a choice between two or more ethical alternatives according to rational criteria of preference has traditionally received the obvious name of the 'rational preferer'. In recent years the rational preferer has been widely discussed, and in general reviled, through arguments which are almost always indebted to Kant's thinking.[1] However, this rational preferer, or a close relative of his, had already appeared in the Scottish empiricist school something like twenty years before the publication of the *Kritik der reinen Vernunft*. The biography of this sought-after and slippery being dates at least from the *Theory of Moral Sentiments* by Adam Smith, whose criterion of moral approbation (to which passing reference was made in the second chapter) represents the critical transcendence of the somewhat naturalist doctrine of moral sense, and consequently it shifts interest in the evaluative question to the field of the rational preferer.

In the introduction to Part II, Section I of the *Theory of Moral Sentiments*, the merit or demerit of an action is located in the harmful or beneficial nature of its consequences, establishing the foundations for an ethical system which was to reach its splendour in the radical utilitarianism of Bentham. Thus, in appearance at least, the question of reward or punishment becomes the corrective element in the balance between benefit and harm necessary to maintain social order. Bad is repaid with bad, good with good. It is merely an empirical question. But a parallel problem appears in that it is necessary to decide which are good actions and which bad, that is, to what extent the consequences of a given action are beneficial or harmful, and (although this would be a separate question) how is acceptance or rejection of this action produced. Once again Smith gives an empirical solution, as he warns in the extensive final note to Chapter V of the same part and section, when he announces that it is a matter of fact and not of law. The principles by which a perfect transcendent being might approve or disapprove of human actions are not what concerns Smith, rather it is those principles used by so weak and imperfect a creature as man in his present condition.

To settle these principles of choice and the form of consent itself, Smith takes as his starting point the critique of the classic solution of the moral sense school which assumes the existence of an automatic mechanism of sanction through a sentiment of a special nature (the moral sense of Hutcheson). That means making a break in empiricism similar to Kant's denouncement of the genetic fallacy, since Smith denies the existence of an automatic link between actor and spectator through sympathy in moral evaluation.[2] But if the classic explanation of moral sense is interrupted, the manner in which acceptance or rejection is carried out must be indicated. Smith defends the rational solution; it is not a case of basing the causes of evaluation on a permanent natural system, but of understanding that a continuous exercise of rational processing is carried out, which, in the light of different situations and through an actor–community interrelation, performs the task of defining the criteria of approval and condemnation in each case. (*Op. cit.* paragraphs 615–617).

This means a new problem is unleashed. The principle of evaluation would receive in the moral sense school a sound automatic foundation, but what now? Might it not occur that between one spectator and another empirical differences in the evaluation of moral actions are produced? The answer to such a simple question has occupied British thinkers for something like two hundred years, because even though it is evident that in the face of doubt about the opposing criterion of the two spectators we have to take heed of the one who is most rational and impartial, it is not at all easy to adequately set out the characteristics this virtuous subject should possess. The history of discussions in this respect is long (and an outline can be found in Broad's writings, 1944–45), but at least it has been able to end on a certain compromise: there are some typical characteristics which the hypothetical spectator must possess if he aspires to the title of rational preferer. The one which has received most attention is that of universalizability, linked above all with Hare. In

Freedom and Reason (1963, Chapter 7, 'Utilitarianism'), Hare announces his intention of endowing utilitarianism with a formal foundation based on the thesis of universalizability, which in fact means going beyond an intrinsic condition of moral judgement, as understood in the prescriptivist school, to the terrain of empirical evaluation which utilitarianism tries to attain. The significance and viability of such a goal have of course been widely debated,[3] but the characteristic of universalizability has been defended, in a broad sense, as necessary to characterize the rational preferer. There is more. Both Firth (1952) and Taylor (1961) have been concerned with the problem, and recently, Muguerza (1977, Chapter VII 'A modo de epílogo: últimas aventuras del preferidor racional') who, following Taylor above all, has systematized the conditions of preference. In a form so précised that it probably does not include the nuances necessary for their meaning to be understood, these conditions can be defined as:

1. Universalizability. Choice must be made free from subjective interests.
2. Information. There must be sufficient information on existing alternatives of action.
3. The rational preferer must not suffer either internal (passions, mental disturbances) or external (coercion, threats, oppression, violence) constraints.

Whoever enjoys such privileges could opt for the position, which as will be seen is vacant, of ideal observer or rational preferer. But Muguerza points out the lamentable error in this system of validation: the ideal nature of the proposal which renders it inoperative. Not much perspicacity is required to understand that even with all the good faith in the world – which would certainly be necessary to overcome the problems posed by the first condition – only an incurable optimist could hope to fulfil the last two points.

But an important question is raised. Where does the difficulty lie? Is it the validation of empiricism which is unfeasible, or, on the contrary, do we have to abandon *any* system of rationalization of moral preference? By definition the second and third conditions are unattainable; it will always be possible to add some other *bit* of information on alternatives which we can in no way take to be complete for certain. It will always be possible to suppose that there are unconscious, or simply unknown contradictions which can limit the liberty of decision. But these are the least of the problems; even when the branches of a hyperbola tend asymptotically to fuse with their tangent in infinity it is technically possible to know which of their points is nearest to contact. Might not a similar mechanism be postulated? Could we not talk of some relative improvements in rational preference capable of rescuing the meaning of the three conditions and thereby define moral progress?

Yes, that would be possible. But we should meet with another, no less serious difficulty. Who defines the asymptotic curve of approximation to the ideal limit? Is it the very rational preferer we are seeking? A historically determined community? Radical empiricism does not seem inclined to lose sleep over the answer, and their help will be of little use to us in the quest. All these questions lack meaning from the strict view of empirical rationality. Rationality is exclusively concerned with the greater or lesser approximation within the asymptote, and not with its form nor the place to which it leads. That is simply not discussed as it is not considered a subject for rational discussion. It is a consequence of the application of the principle of objectivity sufficiently justified from the positivist viewpoint by the misery induced by ignoring it.

That being so, one must necessarily distinguish between (rational) justification of the action as an adequate means of reaching an ultimate goal which exists, and the justification of the goal itself. The rational preferer with all his limitations could take on the first assignment and define, at least, a growing sense of rationality by comparison. And the goal? Well, that is an axiomatic problem, said the empiricist.[4] There is no rationality of ultimate ends in the sense in which the expression is used when we speak of the asymptotic rationality of conduct. If an excessively modest thinker seeks to eliminate this ultimate irrationalism he is avoiding the strict subject of the discussion.

It could be that despite all the empiricist reticence about violation of the principle of objectivity, the moment comes in which modesty has no option but to impose itself on the utility of the dogma (excessively devalued by post-Popperians). There is in fact a recent example which shows the profitability of an excursion through the terrain of rationality of ends as a system of support for the rational preferer, a character who, with a touch of colour, has rapidly taken on new forms in a field almost given up as lost by the empirical cause. I refer of course to Rawls' general theory of justice.

With *A Theory of Justice* (1971) as the culmination of the project repeatedly anticipated in previous articles, Rawls has revolutionized the field of ethical rationalization taken in its widest sense.[5] It had to be so. As in the case of Chomsky, the careful choice of references considered very foreign to the spirit of a vanguard investigator has immediately borne fruit: the goal stated by Rawls is nothing less than the re-elaboration at a higher level of abstraction of the theory of the social contract found in Locke, Kant and Rousseau (*A Theory of Justice*, p. 11). As far as our problem of the rationality of the preferer is concerned, it is Kant who gives Rawls the instruments necessary for his undertaking based on two concepts (certainly quite central in Kantian ethics) of *autonomy* and the *Categorical Imperative*. To see if the stated objective is fulfilled, we have no option but to analyse the way Rawls integrates these concepts into his theory of the *original position*, but perhaps it would be useful to jump ahead and say that these efforts, at times intimately related to Parsons' attempt to construct a general theory of action, do not manage to maintain the high level of formalization desired, nor do they seem to increase the degree of abstraction already used by Kant. Quite the opposite; Rawls' criterion of preference, as I shall try to show, will oblige the use of complementary causal methodology which breaks once again the rationalist asepsis of the analytic school. That is perhaps one of the most important consequences of the Rawlsian formulation, and certainly represents a new way of focusing the relationship between the content of judgement and its context as regards that school. In any case, whether Rawls fulfils his proposal of maintaining a certain level of abstraction is not an essential matter, nor is it worth wasting time over such a subtlety. What is really important is to warn that under the announced project, consequences of such great weight for the very conception of ethics will continue to slip in, and to make the

warning clear we will have to handle some of the concepts used by
Rawls: *rationality*, *veil of ignorance*, and the *original position*.

Rawls uses a concept of instrumental rationality usually found
in sociological theory (as he himself tells us), the only difference
being that he considers a rational person to be one who is not
assailed by feelings of envy. The difference is not significant now,
but the concept of rationality itself is. Rawls is creating a model of
a situation where people confront alternatives, ranking them
according to a criterion of preference which is that of the maximum satisfaction of their desires (1971, p. 143). It is, so to speak,
a militant rationality which is opposed to the idea of intuitive
choice, even assuming that the latter plays some role in ethical
evaluation.

Assuming there is a basis for a rational discussion of ethics, this
requires the existence of another rationality capable of limiting
the intuitionist task of establishing ultimate ends. Thus appear
the two rationalities which we spoke of earlier and which it is
important not to confuse: by the first of them we proclaim the will
of the philosopher to construct rational ethics; by the second we
also put forward the existence of a rationality at the level of the
actors in the social field.[6] And the rationality which acquires most
weight in Rawls' rationalist theory is the second, instrumental
rationality. People engaged in the task of constructing justice as
fairness which will order their lives use the limitations on the role
of intuition, which rest on three criteria:

(a) choosing those principles of justice which would be selected
    in an ideal situation (the original position);
(b) ranking the chosen principles in a sequence of maximum
    conditioned principles (any principle will be maximized if
    those before it on the scale have already been satisfied);
(c) substituting prudential for moral judgement (general abstract
    principles are not discussed but rather limited questions
    which are resolved by somewhat empirical guidelines) (1971,
    pp. 42–44).

It is evident that the task of carrying out these acts of choice and
ranking leads to the criterion of a suspiciously *representative man*
who by establishing himself in the supposed original position

cannot help leaving an odour of the ideal spectator or rational preferer, but as we shall see, Rawls' theory responds to the danger of identification with a certain elegance. First it is worth pointing out the somewhat doubtful question which appears amid the rationality of the actors who suppose themselves to be engaged in the task of constructing justice as fairness. Are we not overestimating their possibilities?

In fact Rawls employs a criterion of rationality which is not used by *the* social theory, but by *a* well known social theory, that of formalism. Even though it is true that the criterion in itself appears in various authors, it is the views formulated from Weber to Parsons and Herkovits which have drawn conclusions that are significant for the social perspective Rawls uses. In synthesis, it is a matter of accepting Robbins' (1932) idea that a human being usually has scant means of facing up to alternative ends, which leads necessarily to the task of choosing. This preference has been sanctioning the concept of *maximization* used by Rawls and implies the basis necessary to define human economic behaviour (or any other behaviour which is concerned with the question of preference) as formally, consciously, and voluntarily governed by criteria of rationality.[7]

But the question is not at all clear. Numerous critiques of the formalist school can be applied to Rawls without hardly any need for correction. Is not the claim to such extended and complete rationalization extremely ingenuous? Even so, it is not the conflict between the substantive and formalist paradigms which we want to bring up here, but the very conflict which appears in the concept of instrumental rationality underlying *A Theory of Justice*. The third step in withdrawal from intuitionist dependency in the task of constructing justice as fairness, the one which talks of *prudential judgement*, implies the reduction of the abstract field of ethical theory by means of guidelines for deliberation, as we suggested earlier. A little paradoxically, Rawls falls back on his own intuition in the battle. The overt practical aim is to achieve a common conception of justice, so that as long as intuitive judgements of priority made by human beings are sufficiently similar, things will continue to function. And if intuitive judgements of priority are different, even at a level which makes agreement impossible? Well, then prudential judgement comes in and establishes which are the most rational points of agreement in the basic structure. Aside from how this choice would be achieved, Rawls indicates later on (paragraph 25, 'The Reasoning Leading to the Two Principles of Justice') the existence of two principles of justice which allow one to suppose that they will be accepted by the system of prudential judgement. These are the principle of equality (basic equalities are demanded for all) and the difference principle (inequalities are allowed as long as they improve the situation of all), defined hierarchically (the equality principle comes first) and under the control, in case of doubt, of the *maximin* rule (alternatives are ranked according to the worst prospects: the alternative whose worst foreseeable result is better than the worst result of the others will be chosen) (1971, pp. 150 ff.).

Now, prudential judgements are impossible to undertake without certain guidelines for the evaluation of alternatives. And those guidelines, which appear in all their magnitude within the Rawlsian principle of the original position, represent a serious handicap for the supposed rationalization of formalism. Rawls explicitly admits that:

Contract theory agrees, then, with utilitarianism in holding that the fundamental principles of justice quite properly depend upon the natural facts about men in society. (. . .) A problem of choice is well defined only if the alternatives are suitably restricted by natural laws and other constraints, and those deciding already have certain inclinations to choose among them. Without a definite structure of this kind the question posed is indeterminate. For this reason we need have no hesitation in making the choice of the principles of justice presuppose a certain theory of social institutions.   (1971, pp. 159–160)

This dependence of social man on natural facts therefore impedes a consideration of the rationality of social actors as something in any event guaranteed. It will be so only when the conditions (ideological above all) of the social structure in which they are immersed permit. In any case 'guidelines for prudent decision' imply a basis of essentially irrational determinants, in the sense that we have no choice but to consider that the decision is at least connected to those pressures and deformations which come from the ideological context in question, and this is variable in terms of historical situations. In consequence, Rawls' ethics needs the support of a causal methodology capable of interpreting social factors which appear in order to sanction the possibility of choice of the original position. Can it be maintained that a Greek contractualist is defending a position of rational logic when he makes the principle of isonomy compatible with that of slavery? Only through a causal perspective broad enough to understand the ideological conception of citizenship and the weight which the gentilique past carries in it. Perhaps in this way we could talk to rational choice, but it is certain to be at the expense of subjecting the concept of rationality to socio-historical relativism.

It might be thought that, after all, the critiques of formalism are not applicable to Rawls, in that he does not direct his energies towards constructing a programme of empirical analysis of social behaviour, but to the simple outline of a heuristic model. Thus for example, Muguerza rejects the totality of Hare's critique of the Rawlsian model (1977, p. 260). However (and even admitting Hare's possible myopia regarding birth ceremonies of novel and promising theories, something which on the other hand usually makes any prominent figure of the establishment threatened so nervous that it justifies the appearance of genuine and pragmatic veils of ignorance) the question to be resolved is what is the significance of formalism *within* the very heuristic model, which of course cannot be rejected under the accusation of employing empirical technical suspicions of reference to 'external facts'. So, Rawls specifically uses the technique of maximization to sustain his model, and situated within this very model are social determinants in the form of guidelines of conduct for prudent preference. The question cannot therefore be posed in terms of whether Rawls is mistaken when he says the peasant tied to a fief does not carry out rational tasks of maximization. What is important is to know the possibilities of preference open to the actors in Rawls' heuristic model irrespective of whether empirically they make use of them. And it is in this sense that the formalist position must be rejected as naïve. The existence of social determinants imposes, on the one hand, a certain level of techniques and strategies of production capable of introducing great variations in the terrain of *input*, that is, of the possible alternatives which are presented as means, and an identical imposition determines closely, in general, the theoretically wide gamut of alternative ends about which the criterion of preference will dictate. But the most serious is that the very criterion in itself can be affected in the heuristic model (in empirical reality it will undoubtedly be affected) by the presence of social determinants. We have no *a priori* guarantee that an impartial judgement can be counted on capable of ensuring the fairness of consequent justice, irrespective of whatever might be the principles employed. So much so that Rawls will be obliged to specify the conditions of his model through the construction of the original position as a theoretical frame of reference. Thus, the problem drifts towards a consideration of the model defined within the very frameword of the original position and the conditions (veil of ignorance) which it imposes, and in such a way that we have to ascertain whether those conditions are compatible with the principle of formalization discussed, that is, whether they permit the development by right (given that empirical development is in fact another matter) of instrumental rationality in its wide form, separate from those principles discussed, or even in limited form, bearing in mind some of the initial principles.

Let us return to the previous example. Let suppose a Greek contractualist is as honourable a member of the original position as any other actor we care to imagine, *providing he fulfils the conditions of the heuristic model*. These of course prohibit the discussion of empirical reasons for what could happen to such a citizen in the public defence of the contract. But they not only permit, but also demand that account be taken of the fact that their evaluation of certain principles (which are going to found their own concept of justice as fairness) will be assisted by some exterior guidelines of conduct, by some 'general facts' capable of ensuring their prudential judgement in excessively abstruse and complicated matters. Let us imagine that equality before the law, together with the presence of slavery, implies a high degree of complication for the actor who is trying to evaluate the meaning of the life and work of a human being. The model prevents us from stating which are the determinants imposed by the 'general facts' of Greek society in, let us say, the 4th Century, but obliges us to take into account the existence of some abstract determinants. The matter for debate could therefore be framed in the following manner: does a general theory of determinants exist, capable of defining a universal system of characterizing prudential judgement in such a way that an empirical case becomes a certain form of the general? If so, Rawls' heuristic model can make use of formalism without problems. But if a general concept of determinism does not exist, if the idea of determinism has no more value than the purely paradigmatic, and the significance and meaning of the ideological determinants of prudential judgement depend in an absolute way on those 'general facts' to the extent one cannot talk of determinism in general because that takes on a very different meaning in different social alternatives, then it is Rawls' heuristic model itself which is of not use to us and for theoretical, not pragmatic reasons. This in essence is Marx's well known accusation of the formalist use of the concept of *property* in a general sense denying the existence of such a concept, and that is the idea of subjugation to the concept of rationality to the socio-historical relativism to which reference was made earlier when talking of the role of reduction of instrumental rationality which the guidelines for prudent decision in Rawls' theory take for granted.

But to accept the preceding argument one would have to pause over the meaning which the original position acquires in the Rawlsian model, if we have not over-estimated the role which the existence of general facts plays in criteria of preference. Straight away Rawls warns that this dependence (admittedly) is less than the utilitarian. But of course. To take the ethical theory of utilitarianism as a point of contrast is no more than, shall we say, a tactical servitude imposed by the empirical context in defence of which Rawls writes. Dependence exists, and it remains to be seen whether its existence threatens the theoretical weaponry of instrumental rationality.

The subject of the original position occupies a considerable part of Rawls' argument; a whole chapter, the third is devoted to analysing its meaning. Moreover, it implies the connection capable of linking neocontractualism with the Kantian theory of contract as a means of combating utilitarian theses, and as the very foundation of new theses. This virtue holds its ground throughout the three parts into which *A Theory of Justice* is divided, always with the recurrent intention of supporting the construction of justice as fairness. "The idea of the original position is to set up a fair procedure so that any principles agreed to will be just" (1971, p. 136).

For that there is nothing better than to define some rigid conditions relative to an initial *status quo* capable of guaranteeing that the principles accepted as the basis of justice are going to have two qualities: rationality and fairness. While the rational decision is an extremely complex problem and forces conditions of fairness and information to be raised to somewhat ideal level, Rawls turns the matter about. He restricts himself to defining a limited situation, in such a way that the data we are using for a rational decision are limited, and as such can be taken for granted and weighed up without pretensions of transcendentalism. It might be thought that in this way little ground is gained in a field as complicated as that of jurisprudence, but Rawls shows himself to be extremely (and deceptively) modest in his pretensions. For him it will be sufficient to find a few premises, however weak they might be, capable of leading to certain conclusions acceptable as rational and impartial. These conclusions will manage to impose, jointly, the limits between which the principles of justice as fairness will be ordered. "The ideal outcome would be that these conditions determine a unique set of principles;" says Rawls, "but I shall be satisfied if they suffice to rank the main traditional conceptions of social justice". We should not be deceived by the impression of modesty. If we have discovered how to rank the principles of justice in a rational way, we will find ourselves in a situation where in fact we have defined through rational criteria what is needed in order to proceed. As I said in the introduction to the theme of the preferer, that is precisely the task underlying decision theory, and so from the chosen ends it can pass to a second plane of validation of means - much less embarassing. If one gives in to the verifying arts of positivist science, or if an *ad hoc* instrumental reason is postulated, the problem of the rationality of means is by far inferior in complexity to the thorny one of the rationality of the ends. And Rawls tackles the latter. But it makes no sense to repeat what will be discussed later on. First we must examine the reasons on which the Rawlsian criterion of final justification is founded.

The original position is based, as we have seen, on a starting point capable of limiting the field of preference. It means that men in the process of accepting some common criteria have to be stripped of those contingencies which usually lead them to different situations from which, in consequence, unequal choices are made. Inequality, for Rawls, comes fundamentally from the tendency to exploit different situations for one's own benefit, and the characteristics of the original position will therefore have to be those which impede the development of this selfishness. The appropriate situation for the rational decision turns out to be opposite to the one sanctioned by the classic doctrine of the rational preferer, that is, it is a radical reduction in the level of knowledge in which actors move. Rawls calls the model artefact capable of guaranteeing the purity of ignorance the veil of ignorance: "Now in order to do this I assume that the parties are situated behind a veil of ignorance. They do not know how the various alternatives will affect their own particular case and they are obliged to evaluate principles solely on the basis of general considerations" (1971, pp. 136–137). It is a very difficult imposition to accept as empirically present, but it is worth remembering that nothing less than the logic of the heuristic model of justice of Rawls is under discussion. Whether it is valid, or rather useful to explain the degree of fairness of the justice in a particular community is beside the point; we are at present only interested in weighing up the operative significance of postulating the veil of ignorance for Rawls' theory of justice. The rejection of the mechanism of the rational preferer is due, let us recall, to difficulties of theoretical and indefinite extension of the conditions of universalizability, information, and freedom, which rapidly take the path of the asymptote, but in this case it is not a definite extension, but a decisive limitation which defines the ideal situation. And the limitations are easily specified.

As a consequence Rawls proclaims victory, and this time with few pretensions of modesty since from his point of view what the original position achieves is to support a rationality concordant with the Kantian Categorical Imperative no less. If the latter is established on the basis of the concept of autonomy of the human person, it is the veil of ignorance which ensures the absence of heteronomous contaminants capable of distorting the dictates of free and rational nature (all of paragraph 40 is devoted to establishing the bridge between the Categorical Imperative and the neocontractualist principles of justice as fairness). Of course, if the veil of ignorance is able to place actors in a situation of equality and independence, the conditions for the development of rational decisions corresponding to those postulated by Kant in his formulation of the Categorical Imperative, it remains to be seen whether we are overvaluing once more the possibilities of the model. Rawls seems to do so excessively when he takes advantage of the Kantian concept of autonomy to take the final step from the perspective of the heuristic model to that of the empirical development of rational choice. Although it is a little long, it is worth quoting the passage:

> Assuming, then, that the reasoning in favor of the principles of justice is correct, we can say that when persons act on these principles they are acting in accordance with principles that they would choose as rational and independent persons in an original position of equality. The principles of their actions do not depend upon social or natural contingencies, nor do they reflect the bias of the particulars of their plans of life or the aspirations that motivate them. By acting from these principles persons express their nature as free and equal rational beings subject to the general conditions of human life. For to express one's nature as a being of a particular kind is to act on the principles that would be chosen if this nature were the decisive determining element. Of course, the choice of the parties in the original position is subject to the restrictions of that situation. But when we knowingly act on the principles of justice in the ordinary course of events, we deliberately assume the limitations of the original position. One reason for doing this, for persons who can do so and want to, is to express their nature as free and rational beings. (1971, pp. 252–253)

There are at least three implications of interest in this paragraph from Rawls: (a) the principles governing actions by those in the original position do not depend on social or natural contingencies; (b) human nature is of a free and rational being, subject to the general conditions of human life; (c) restrictions on the original position are assumed voluntarily in the ordinary course of events, as an expression of free and rational nature. Other implications of the paragraph, like the restriction of particular interests, already appeared as assumptions in the original position. But, how can we claim in these conditions that the principles governing the action of citizens immersed in the original position do not depend on natural and social contingencies? Both contin-

gencies, by definition, form part of the mechanism which transforms the veil of ignorance into a partial barrier. Although actors
lack information about their own positions in the social framework, they not only necessarily possess information about the
general conditions of their social system, but also find themselves
subject to consequent determinants. One leads to the other, and
the intention of offering a social perspective of elaboration on the
Kantian imperative imposes these obligations.

If Rawls is able to avoid the arguments which invalidate the
mechanism of the rational preferer, he does so by accepting
general facts as a means of particularizing the task of choosing. It
is not a case of a preference of absolutely abstract ends, methodologically applicable even to transcendent beings, as Rawls
assures us when detailing the differences which exist between
justice as fairness and the Categorical Imperative, but of an
earthly question of establishing criteria of choice with regard to
some alternatives which are related to certain impositions of
nature and the social structure. But, that same guarantee of
escape becomes an obstacle for the use of the Kantian sense of
autonomy. The theory of justice as fairness thus teeters between
a formalization which cannot be any higher than that permitted
by the 'general facts' as determinants, nor so low that it distorts
the rationality of the criterion of the model. While general facts
contain these distorting paths with great ease, Rawls' solution
quickly drifts towards the course of the rational preferer if it tries
to rescue its high level of formalization. So, in fact we are back to
where we started. Either we face the existence of certain biological and historical determinants and try to give an account of how
it is possible to save rational preference after all, or we end up
accepting formal dualism and we have to confront the more than
serious doubts about how the mechanism of preference could
function – unless we maintain the idea of changing the whole
panorama and claim that, when all is said and done, ethics has
little to do with rationality, and what is true or false in a moral
sense are very different terms to those invoked in scientific
methodology (somewhat distorted by post-Popperians). In that
sense it is necessary to pay heed to an author, Chaim Perelman,
who has proposed what seems in fact to become a 'third way' as
opposed to the alternative between cause and reason, invoking

measures expressly related to Aristotle's dialectics; a way which
the author designates with the sonorous name of the *nouvelle
rhétorique*.[8]

Against the simple acceptance of the 'rational' approach to ethics, voices as numerous as they are well known have been raised,
and the Marx-Nietzsche-Darwin axis is a very obvious example.
It is clear that we would be committing a clumsy sin of simplifica-
tion if we were to think that it is an easy triangle to reconcile,
even if we were to reduce the rationalism-irrationalism opposition in ethics to the terrain of the meta-theory. Neither Marx, nor
Darwin, nor Nietzsche claimed of course to establish an irrational
theory of ethics, but a very rational explanation of the manner in
which human actions (in ways quite different according to each
author) fall within the sphere of an irrational adherence, in the
sense of determined by non-rational compulsions, to so-called
ethical truths. That is so obvious that it is hardly worth mentioning, but it is worth recalling the way in which irrationality is
brought into the field of ethics by those authors, inasmuch as it is
identifiable with the classic alternative line of the justification of
moral conduct apart from the logical elements of discourse involved in it.

   With Perelman the matter changes, or so it seems. It is not a
case of arguing about the presence of rational elements (analysable with the tools of the logic of language) or irrational ones
(which claim the interest in moral questions of those within fields
in principle distant from philosophy, from political economy
across to biology) but that human action would be, perhaps not
exclusively, but alternatively analysable from a field in which the
respective relations between ethics and reason, between axiology
and logic, are profoundly modified. Because in Perelman's studies
of rhetoric, the central matter of action does not turn out to be
either rational or irrational, but something different. Neither is it
a case of play on words; the area of analysis which Perelman is
marking out does not belong to the classically irrational alternative, but its very language and relation to logic are submitted to
the scalpel. But neither can the Belgian's approach be so easily
located within the matter of the rationality of moral language as it
is expressly denied in the analytic sense. The term 'arationality'

could perhaps express this intermediate zone, if it were not for Perelman's interest in rationalism.

Perelman's principal theses seek to mark the differences between the rational system (that of formal logic) and the rhetorical system regarding the possible use of the latter as an instrument for establishing a practical validation, that is to get support for moral theses different, of course, from that which would be obtained from logical resources. The task is not easy because it includes an operation of change in the criterion of rationality. If this is identified with the path taken by formal logic without further ado, all alternatives would obviously become irrational: in that sense one talks of the irrationality of evaluative discourse. But if we abandon this perspective, if we deny the 'fine distinction' between formal logic and virgin jungle, an intermediate ground appears in which even if we cannot justify with formal criteria, we can always analyse, classify and draw conclusions regarding the primordial question of how adherence to ideas is achieved. This is of course the field of rhetoric, and one would have to grant the right to life to such a theoretical-methodological conglomeration as long as we take as complete the task first assumed, that rhetoric is distinguished from formal logic, and simultaneously offers a field of study according to conditions generally used in philosophical analysis.

Let us now see how such a thing can be achieved. In *Logique Juridique. Nouvelle rhétorique* (1976), Perelman established some general lines about the relations between formal logic and rhetoric, which were to be repeated in *L'empire rhétorique* (1977) and give sufficient leads about what can be expected from this new methodology. Perelman's four observations in this respect are:

1. Rhetoric tries to persuade through discourse. There is no rhetoric when experience is resorted to in order to obtain adherence to a statement, nor through the use of violence or caresses, unless situated in the linguistic field (threats and promises).
2. Rhetoric and formal logic are two different things.
3. Adherence to a thesis can be of variable intensity.
4. Rhetoric refers to adherence, not to truth (1976, paragraph 51).

The difference between rhetoric and formal logic is due to the vague, disordered, and multiple character of language, something which formal logic ignores when it claims to be based on the Cartesian idea of clear, distinct ideas. In the field of rhetoric there is nothing similar to the rationalist 'truth' which could be reached through irrefutable evidence. It is necessary to argue to achieve adherence, which, by the third thesis, will be a variable thing. It is however important not to confuse the sense of rhetorical adherence. When Konrad Lorenz says that people allow themselves to be convinced through such irrational mechanisms as the sound of a hymn or the fluttering of a flag he is proposing something suspiciously similar to what Perelman states, but if we try to analyse the use of hymns and flags to achieve certain goals established in the moral order from the perspective of ethology, we will have radically diverged from Perelman's perspective to enter the psychology of masses, political sociology and ethology. I have maintained that that is necessary in order to have the possibility of analysing moral phenomena with some hope of success, but it seems that Perelman does not accept such liberal extensions of the ethical theory. The interest of rhetoric is 'philosophical' in a restricted sense of the term, and refers exclusively to things like argument, controversy, and dialogue. It is not that dialogue cannot be taken into account as an element of ethological interest for example, but that once again we have entered into the question of perspective. Perelman's theses must make sense from their own standpoint and not by turning to external explanations. If we agree to say it in a pedantic way we would have to conclude there is a need for discourse to contain its own legitimation.

But that is a very serious matter. In situating rhetoric's field of action in discourse itself, in explicitly throwing out incursions into the extralinguistic field, Perelman is, on the one hand, laying the foundations for a paradoxical rational faith, which he will complete later on, and more importantly for our purpose, confirming suspicions about the internal legitimation of discourse. He is trying to propose something new to us, something unconnected with the traditional attack on axiology by the so-called positivist sciences or by the theorists of ethical irrationalism. By obtaining persuasion (the ultimate goal of rhetoric) through exclusively

linguistic elements, we cannot use the support of the categorical
tools of sociology or psychology in our analysis, which is equivalent to admitting that discourse alone contains all the significant
factors. There will clearly be an escape valve which allows analysis
of how elements of discourse are articulated and completed, but
we cannot take this idea to its final conclusions: if we suppose that
adherence to a given axiological environment (which corresponds
for example to the transcendent nature of private property) is
achieved thanks to linguistic development obtained within a given
community with an adequate sociopolitical system (and we imagine that this correspondence leads to bourgeois society pro-
tected by parliamentary democracy) the relations between rhetoric
and the system will have to be limited to the use of linguistic
expressions. We cannot assume that the set of ethical values is
maintained through police charges, shall we say, because in that
way we are going beyond the rules of the game. One could
threaten, it is true, but without serious political or sociological
consequences.

Such a matter raises grave doubts about the autonomy of
discourse. Let us imagine a western parliament in which certain
citizens with the role of members of parliament try to obtain
support of minds for theses presented for their assent.[9] The minds
in this case are the members of the chamber (and eventually the
whole country), and the theses presented are those referring to
the need for a law restricting civic liberties. This is an especially
suitable example in a theory of argument, as it defines an
audience and a *presence* very much in accordance, as we shall see
later on, with the general sense of Perelman's theses. The weapon
of conviction cannot be formal logic of course, so it would be
useless to argue in terms of the truth; syllogisms will only be used
as an extra force to complete the discourse which seeks to
persuade. And the extralinguistic environment? Here lies the
problem. If a member of parliament belonging to a small party
with scant national support and negligible relations with strong
groups capable of using street violence makes a speech full of
threats he will obtain a very low degree of persuasion compared
to the other citizen who occupies, let us say, the position of
minister of the interior and who raises the possibility of a strike by
the forces of law and order, or the reverse, a veiled threat of

armed intervention by the police and army. Both threats belong to the field of linguistic expression; both use a series of concepts equally coined by the system, but in no way could they lead to similar rhetorical results. Either we are not now doing something which can be analysed from the perspective of rhetoric, which is inadmissible because we have not left the field of argument of linguistic expression which *includes* threats and promises, or extralinguistic elements have greater weight than Perelman claims.

Is this not a borderline case, one of those uncomfortable frontiers which it is best to ignore? It seems not. The persuasive development of discourse commonly includes relationships with the extralinguistic environment which cannot be ignored. On the contrary, it will be at a later moment that linguistic autonomy is produced, when expressions become fossilized, (which Emilio Lledó so rightly emphasizes) breaking their links with the social environment. The reference for example to *masonic collusions* or *obsolete liberalism* becomes, at least in the Spanish language, closed and coin a sealed concept, impermeable to its primitive sense, thus claiming the exclusively linguistic field (at least in principle). We would then be in a wretched field for rhetorical analysis; we would have passed into a field of play in which adherence and persuasion are produced automatically, acritically, leaving very little or no room for the rationalization of the process.

Let us see however if it is possible to find a chink in scepticism. Let us go to the third of Perelman's observations in case a way out can be found. Let us recall that adherence to a thesis can be of variable intensity.

This statement leads to the establishment of differences between speeches about real facts and speeches about values (paragraph 54). The truth, referring to the empirical world, does not assume either adherence or discussion: there is no reason at all to prefer the false one. On the contrary, it is indeed possible under given conditions to prefer one value to another while adherence is not contradictory; one value is opposed to another, not a 'false value' (although rhetorical propaganda can use the concept of false value) and the only thing that has to be achieved on rhetorical grounds is a type of change in intensity of adherence).

Now it is not a case of carrying on a discussion about the sense of deontology and its possible connections with, shall we say, the

Categorical Imperative, but of understanding that Perelman's statement necessarily leads to an image of rhetoric in which the system of values and its different levels of adherence lack any validity without a frame of reference. By introducing comparative terms of intensity we are obviously going to need a system of co-ordinates. And so we arrive at what could be classed the heavy artillery of the argument theory, with the concepts of *commonplace*, *audience* and *presence*.

For Perelman the notion of audience is central in rhetoric (paragraph 52), given that discourse is only effective if it is adapted to the audience it tries to persuade or convince. Let us be clear, this adaptation does not mean transgressing once again the frontiers of linguistics, unless we fall into circumstances such as those mentioned earlier; discourse can mould itself perfectly to a changing audience and it could be thought that this is to a certain extent even an automatic change, as it is likely that a great part of linguistic inheritance stems eventually from the audience to whom the discourse itself is directed. There is no reason for scandal in supposing that rhetoric works by feedback. But it does pose a curious question. Let us imagine an attempt to persuade about a certain thesis; by definition rhetorical development will bear a relationship to the actual audience to whom techniques of adherence would be directed. Let us now suppose that the thesis to be sold stays the same, but the audience changes. It will be necessary to adapt the structure of the discourse to the new conditions, since we assume that the audience is not the same. And what if we were to maintain the original discourse and in spite of that achieve adherence at a tolerable degree of intensity? Then quite simply we would have achieved a set of arguments which had a special characteristic, that is to say, with a double capacity of conviction. It would clearly be a very useful discourse. And if those arguments can have a multiple use, and thus be accepted by a universal audience? We would be dealing with *commonplaces*, with Aristotelian dialectical rhetoric. For Perelman, *commonplaces* play a role in argument analogous to that of axioms in a formal system (paragraph 58), although they maintain their ambiguous and changing character relative to the environment: it is enough that there is a degree of accord with the starting point, with general matters such as 'freedom is preferable to slavery', although in fact the meaning of free and slave may vary from one society to another.

The significance of this thesis is, then, immense. If we accept that through the concepts of *audience* and *commonplace* it is possible to lay the foundations for the development of a task of persuasion, we will have overcome the barrier of irrationality which originally separated evaluative judgements from those of reality. Practical reason finds a field of activity which can even be extended to the positivist field itself at the moment, which seems inevitable, when the system of validation of scientific thought cracks. For Perelman it is the triumph of rationalism over a narrow empiricism, certainly useless for approaching the field of ethics, or over a mistaken metaphysics. Perelman warns of the errors of ontology, which attempts to find an objective foundation for values and norms capable of eliminating ambiguity and introducing the concept of truth into axiology (paragraph 56), and of the uselessness of converting practical problems into theoretical ones thanks to the inclusion of a science of conduct. That is to say, he rejects *both* alternatives regarding the relation between morals and reason, the rationalist and the empiricist, on behalf of a new rationalism.

The attack on empirical excesses is based on the origin of the *commonplace*. And where might such an interesting element originate? Allow me to transcribe the paragraph in question:

If it is indisputable that all argument presupposes the audience's adherence to certain theses and certain prior opinions, empirical epistemology which attempts to derive all our ideas from experience must be rejected as it forgets that, besides experience, whose role in controlling and correcting our ideas is undeniable, these ideas constitute a prior element, transmitted by education and tradition, and they need the existence of a common language as synthesis and symbol of a culture (. . .) Learning a language also represents adherence to values which this language brings with it, explicitly or implicitly, to the theories which have left their mark on it and to the classifications which underlie the use of terms. (paragraph 55)

Yes, it is an attack on empiricism, an empiricism which is indeed naïve and primitive, which any Skinnerian would be able to refute with certain elegance. This view is not at all original. It is not Chomsky who is shown up by the relentless and destructive work,

but Sapir, Whorf, or even Humboldt. It is obvious why: a strong
rationalism in the Chomskyan sense would oblige those *commonplaces* to be situated beyond mere linguistic inheritance,
within deep grammar. It is a question which has already been
dealt with and I am not going to repeat in here. In consequence
this characteristic of centrist theory is maintained in Perelman's
scheme, and rationalism receives a corresponding corrective:

While the axioms of a formal system make abstractions from the context (. . .)
argument is necessarily inserted in a psychosocial context which cannot be
entirely separated from underlying forces (. . .) This perspective cannot attract
those who ignore everything about argument and regard practical reason on the
model of theoretical reason, and preferably from a formal reason. This approach has led a large number of philosophers to the search of some primary
moral principles and to present them as evident, or at least as not controversial
in an environment, given that such a point seems to impose itself in a certain
ideological climate. However, it is sufficient to point out that such principles are
very numerous and seem at first sight to be incompatible, although we could
make an effort to reconcile them, to realise how vague they are and the extent to
which their validity is doubtful. (paragraph 62)

Now it is Perelman who is guilty of empirical prejudice. A
theoretical thesis, like for example the Categorical Imperative,
cannot be combated by making use of experience which shows us
how variable and contradictory are the 'first principles' of morals.
The only thing which that ensures is that we have been mistaken
in marking out the so-called imperatives. If 'dignity' and 'freedom' are incompatible in a certain social environment, the fault
does not lie with the Kantian thesis, but with our way of understanding it and expertise in detecting authentic universals. But
Perelman needs to define the claim of rationality excluding uni-
versalizations, which would convert the field of gradual and
variable adherence into that of formal confirmation of truth and
lies, or they would finally get positivist influence, a 'science of
conduct' capable of establishing the bridge. As has already been
seen, by developing Chomsky's theses, a system of causal foundation of that type appears.

Thus we reach a new *impasse*. Discourse tries to become
autonomous by virtue of the conditions established by the audience which ultimately manage to establish sufficient common-

places to construct the new rhetoric as a system of rationalization. Reason and common sense for Perelman have indisputable links (paragraph 58). But common sense includes giving up a great deal of autonomy of discourse, because it relies totally on an environment able not only to generate linguistic elements, which is in principle as logical as it is acceptable, but also on modifying the task of adherence, introducing extra-axiological (to put it that way) disturbances, or even better, of changing axiology through references and interlinkings within which can also be established an analysis of the opposite type, that of the modification of the non-conceptual environment by rhetorical means.

Thus Perelman's theses come very close to the problems which lead to certain hesitations in the Rawlsian theory of justice. Between the relativism of discourse and its claim to autonomy there is a distance difficult to fill. In fact, it might be thought that all ethical theory today is moving uncomfortably along a narrow path bordered by these two alternatives.

CHAPTER 6

THE GAMMA-MORAL LEVEL: GENES AND TYRANTS

Do you not know, he said, that some cities are ruled by a despot, others by the people, and others again by the aristocracy? – Of course.

And this element has the power and rules in every city? – Certainly.

Yes, and each government makes laws to its own advantage: democracy makes democratic laws, a despotism makes despotic laws, and so with the others, and when they have made these laws they declare this to be just for the subjects, that is, their own advantage, and they punish him who transgresses the laws as lawless and unjust. This then, my good man, is what I say justice is, the same in all cities, the advantage of the established government, and correct reasoning will conclude that the just is the same everywhere, the advantage of the stronger.

Plato, *The Republic*

The gamma-moral level is made up of the norms of cities: those rules of conduct which are constructed in historical conditions within different human groups. Thrasymacus shows us that the multiple, even contradictory, nature of forms of organisation suggests moral relativism. But he finally includes an idea about the unique character of what is just – it is what suits the strongest.

Thrasymacus' words could be re-read from a parallel perspective by considering that the functional character of ethical conduct is capable of imposing gamma-moral contents which suit the strongest line of biological adaptation. It is a thesis which has been held in sociobiology, and which however seems to be easily refutable without going further than referring to the enormous diversity of moral codes. But determinist theses are too sophisticated to be fought through so clumsy a methodological device as recourse to the empirical datum. Let us modify the type of question we are going to ask a little: in what way *might* normative elements on the gamma-moral level depend on the biological substrata of individuals who maintain and respect (within certain limits) positive rules? Or put differently, is there any type of theoretical formulation able to render comprehensible the way such dependence comes about?

Sociobiology offers an answer based on the evidence obtained in population genetics, an answer which perhaps does not lead to the strict determinism which would be illustrated through a mechanical dependence, but which at least contains the outline of tendencies towards the maintenance of a certain type of practical moral rule suited to the ecological conditions of the environment. Given that the environment is diverse, not homogeneous, empirically confirmable moral diversity itself would cease to be an obstacle.

Of course the existence of morals as a phenomenon capable of directing human conduct with relative force obliges us to consider that moral rules cannot be indifferent to the fact of adaptation to the environment. Although we might maintain the most radical thesis referring to the separation between rules and nature, postulating the existence of an absolute and unbreachable rift which prevents all types of natural contamination in the formulation of positive ethical norms, it would be impossible to claim the reverse, that the influence of moral norms in phenomena of the natural order is equally unfeasible. In fact the relation between one environment and the other is close (analogously, of course) to the traditionally thorny matter of relations between inheritance and environment in the field of genetics. The central dogma of molecular genetics formulated by the discoverers of the double helix structure of nucleic acids has contributed to the location of relations between inheritance and environment in their proper place. Desoxyribonucleic acid, DNA, contains coded genetic information which is transcribed in the molecule of another nucleic acid, RNA, and is finally translated, giving rise to protein synthesis. It is a process in which genetic information circulates only in the direction DNA-RNA-protein; this flow of information cannot be reversed. Biologists like Jacques Monod are adamant on this point: it is not that empirical proof of a reversal in the flow of information does not exist, nor even that there are conceivable mechanisms by which the characteristics of the protein could communicate any information to DNA (1970, p. 123).[1] And this unquestionable thesis serves to disqualify the Lamarckian model of the transmission of acquired characteristics by inheritance (of events which modify proteins in the course of individuals' lives).

But it would be an error to deduce from the central dogma of molecular genetics absolute impermeability between inherited matter and the environment. All that has been done is to substitute sounder models for the classic interpretations of relations between environment and genetic code (Lamarck or Weismann's theories of inheritance).

DNA contains information in principle separate from all environmental circumstance, which can vary in the process of genetic duplication in a totally random, undirected way. But the environment which surrounds the mutant individual is not indifferent to the process of change, far from it. It represents, on the contrary, a source of determinants, because the environment is able to 'choose' among the multiple variants to which the change in information contained in DNA, or errors of transcription or translation, can give rise, allowing some mutant forms to progress and become new and prosperous sequences of DNA over generations, or preventing the individual carrying new messages from managing in turn to transmit them, through incompatibility (or for mere inferiority relative to the rate of reproduction of individuals).

In the case of relations between nature and the moral, not genetic code, they are so different that it is dangerous to speak of analogies. But these exist, at least in one sense; that of permeability. Even where we demand that the moral code be independent of natural circumstances, and respond to the free exercise of human will more or less moulded by historical events, we find ourselves in a situation somewhat similar to that of geneticists. If positive moral formulations vary, in principle, in a way outside the scope of the ecological environment (which, from the perspective of the environment could be expressed by saying that they would vary 'by chance'), in no way can we draw the conclusion that they are foreign to it. At the very least links would be established by the fact that the presence of some given moral norms in human groups is going to have serious consequences for the way men in those groups respond to stimuli coming from the ecological niche in which they find themselves. And if that is so, the environment will without doubt eliminate as incompatible all those moral tendencies which reward conduct highly negative for the task of adaptation, through the radical route of eliminating the very groups who maintain them.

Of course, this is a limited and extremely clumsy postulate, because, among other things, it is improbable that history runs along these channels. But if we consider that man, for special motives, can make his links with the environment very elastic (within certain limits) and doesn't arrive at extreme situations of unfortunate consequences without first having reorganized his particular moral vision of the world, we will have no choice but to accept that the ecological niche is able to impose extreme determinants on the type of moral rules which are present in human groups. If we deny this, radically invoking the principle of autonomy which appears in the second aspect examined on the beta-moral level, the disappearance of the group ceases to be a ridiculous hypothesis. Taboos on eating the meat of a particular animal cannot be maintained if there is no other way of getting proteins, or better put, they can, but for very little time.[2]

But the philosophical significance of determinants from the environment and the genetic code on the gamma-moral level would be slight if it were only related to the idea that what cannot be done cannot be done. At the moment it even seems that we have just supported the opposite thesis, by considering that it is the gamma-moral level norms which impose severe determinants on what can happen later in the framework of adaptation to the environment. In individual considerations, it could be so, but speaking in global terms it would be necessary to return to the analogy about the correlations between genetic code and environment. The hierarchical relation could be inverted without further ado, since although the source of diversities is considered fundamentally to be by chance, it is the environment which introduces into the equation of incompatibilities the real independent variables. The environment has at least some elements which can be manipulated from a human scale, and as we shall see later, this fact is of great significance. But there are also ecological conditions which we cannot modify and in that sense they are independent, autonomous, data from our viewpoint. The presence of these conditions means that however much moral norms can vary in origin, they will necessarily be ranked by the environment in the task of adaptation. And that in essence is the argument underlying sociobiology theses. There is no need to pose matters in terms of all or nothing, considering the environment only in extreme conditions. To accept this means also having to put forward the presence of environmental circumstances compatible to a greater or lesser extent with certain values, perhaps not so basic as to mean the permanence or disappearance of the group, but certainly sufficiently strong as to modify in one sense or another the direction of normative codes.

The significance of these determinants, and their relation to biological norms has been formulated making use of the conceptual framework of population genetics, and more specifically the theory of the $r - K$ continuum.

In the reproduction of the species we can in general talk of two types of opposing strategy. There are species where individuals reproduce by means of huge quantities of offspring, to whom hardly any care is given, or conversely, species formed by beings with few descendents who receive considerable attention. For their part, environmental conditions of selective pressure can also be understood by means of a radical alternative: that of a regular and constant availability of resources (in the final instance, food) or the opposite condition of different variable and irregular periods in which large lapses of time pass with absolute scarcity and brief interruptions in which resources suddenly abound. Of course, all these polarized extremes are only theories which can give way to a multitude of intermediate situations in the reality of nature, but the interpretative model is more simple if we limit the possibilities to the two extremes in each case.

Between selective pressure and the strategy of reproduction there are interesting correlations which sociobiology formulates through two conditions of analysis, two factors which express the possibilities of population growth in terms of the quantity of resources available, factor r, or intrinsic rate of natural growth, and factor K, or the carrying capacity of the environment. If we restrict the model in the sense of holding that there are no migratory phenomena, factor r (or Malthusian parameter) is usually considered an expression of the difference which exists between average birth and death rate of individuals over a unit of time. If that difference remained constant over time, population growth would take off exponentially. This is not so because birth and death rates are not independent variables, but functions of the total population existing through a relation which is influenced by conditions of the environment and the resources which it can provide. So, basically, we can maintain that the limit on growth is represented by the capacity of the environment to provide resources, of if preferred, to bear that maximum population which would appear when birth and death rates are balanced, that is to say, when the Malthusian parameter r is equal to zero. The maximum theoretically calculable population in that form is precisely factor K, the carrying capacity of the environment.[3]

So, reproductive strategies of populations can be related to the two factors r and K simply by taking account of what type of variations in the rhythm of growth would be favoured by the environment. A species which maintains a very high number of births with scarce attention to new beings would tend to reach the maximum possible population in a short period, although with a large number of sacrifices among the multitude of beings born as the population continues to grow. This type of almost exponential growth is called strategy r. At the other extreme, a species which perpetuates itself through few offspring, to whom it devotes great care, will reach its maximum more slowly, but will maintain itself near that point without difficulties and with little sacrifice of 'surplus' individuals. It will be a species relying on strategy K.

Environment can selectively favour the adoption of one strategy or the other. Wilson indicates the covering of weeds in a new clearing in the wood, the muddy surface of new reedbeds of a river, or the bottom of rain puddles as habitats suitable for an 'opportunist species' (with strategy r). Here a species can only prosper through speed in finding and making use of resources (which implies rapid reproduction towards the maximum possible), and so genotypes with a tendency towards an elevated r will be favoured. Ideal habitats for a 'stable species' (with strategy K) would be an old wood, the wall of a cavern or the interior of a coral reef. Species which thrive do so by extracting energy in competition with other species who between them almost exhaust the resources, and those who prosper will be those who produce individuals highly selected for a complex task, even when growth is achieved slowly, which leads to the favouring of genotypes with the K tendency. As presented, the model is very simple, because there are factors determining the size of the group of which I am not taking account and moreover since there is no reason for r and K strategies always to be polarized alternatives, but the reader interested in those details can find the relevant details in more suitable places. For our purposes we can start from this reductionist overview, or even mix it with some form of generalization. So, we could assume that the environment propitious for an r strategy coincides with that found in extreme climates, with a brief spring and summer period in which there is an explosion of life, and in consequence an availability of food, while the habitat suitable for K strategies could be that of mild climates with few fluctuations where there are regular averages of resources. Thus, we would avoid having to project onto the behaviour of human groups things so exotic to man as the bottom of rain puddles, the interior of coral reefs, the walls of caverns or the mud of reed-beds. On the contrary, since both extreme and moderate climates are habitable for us, one could ask whether the alternative between r and K strategies is significant for the human species and above all whether it has anything to do with the question of the formulation of classifiable, positive, principles in the delta-moral level.

For sociobiology mankind is an example of a species selected through K strategies. It reproduces through very few children and devotes more time to caring for them than does any other known species, and slowly draws close to the maximum possible limits of population tolerated by the environment, but with the result that groups tend in general to find themselves relatively close to factor K. However, the advantages of the model of the r and K continuum have encouraged sociobiologists to exploit it as a source of classifications which are somewhat removed from the initial aim and define *cultural r* and *K* strategies in a sense closer to that which was held by mere demographic variation. From the value which sociobiology grants culture this is not an excessively heterodox operation, since cultural responses to environmental stimuli could be understood simply as adaptation factors of identical standing to that which, for example, the genetically fixed birth rate might have. If a group systematically practises infanticide for culturally directed motives the difference is not going to be very great from the consequences which would result from a decrease in the capacity to engender numerous children. As we shall see later on, it does indeed allow distinctions to be postulated, but not for this specific motive.

So the $r - K$ continuum broadened from this new perspective can express a theory about the possibilities of ranking moral rules according to the impositions of the environment. r groups would be those continually warring with their neighbours, who undertake territorial expansion by force (or even in a pacific manner, understanding by this competition with non-human species) of which an example could be Viking society or any of the states which practised colonialism. Those with self-sufficiency, centripetal behaviour, who rely on great masses of population and are near the state of complete exhaustion of the resources of their surroundings, like traditional Chinese villages, would constitute K groups. And the moral values of one or the other, according to sociobiologists, would differ remarkably, ranging from selfishness, appreciation of numerous children and the praise of youth as ethical components of r groups, to altruism, pacific solution of conflicts, nepotism, and the cult of old age, which are predominant values in K groups.[4] The way in which the hierarchy is presented is quite obvious: it is much more plausible to explain the cult of the values of boldness and daring through ecological determinism than the reverse, since the environment contains elements which not even the culture itself can modify. However, the previous statement should be qualified in a way which sociobiologists do not fully appreciate. I refer to the historical tendency towards progress, understood as the accumulation of technical means.

One of the most outstanding members of the neo-evolutionary school of anthropology, Leslie White, warned as early as 1949

that the balance between energy and culture should be considered as a process of evolution subject to natural limits. In any
cultural situation there exist, for White, three factors which relate
the environment to the production of goods, and these are the
quantity of per capita energy obtained in a given time, the
efficiency of the technical means available to the group to make
use of the energy obtained, and the total amount of goods and
services produced by the group in that same lapse of time.
Assuming that other factors in the ecological niche, like the
fertility of the soil or livestock, the average rainfall, or others of
this type stay constant, the basic law of cultural evolution formulated by White states that culture can only progress as one of the
following two things (or both together) increases: the quantity of
available energy or the efficiency of technical means in making
use of it (1949, p. 341). Obviously this is a law which must be
understood in broad terms with regard to the concept of culture it
contains. Some cultural aspects could indeed evolve in pro-
gression even though there might be an energy freeze within the
resources available to the group, but in terms of growth in the
volume of population and its establishment in a given habitat
White's equation is extremely thought-provoking, above all with
regard to one specific matter: what White is suggesting is the
possibility open to human groups of modifying factor K.

In the line of argument around the equations expressing the
carrying capacity of the environment, the straight lines representing the variation in birth and death rates, connected to the
coefficients which indicate their slope, are taken as dependent
variables of the characteristics of the habitat.[5] But the very
capacity of the environment, that is, the totality of its resources,
is a factor which does not appear anywhere. Now we have no
choice but to take it into account. A phenomenon like that which
Faustino Cordón calls 'autotrofismo', that is, the incorporation of
a new foodstuff hitherto inaccessible into the relevant environment introduces considerable modifications to factor K. And the
human species adopts this type of strategy, perhaps not in a
continuous, but in an important way if we take long periods of
time into account. The cultural evolution which White defines is
in fact only a basic modification of the K factor and the history of
humanity coincides with that of the alternatives in the task of

extending the capacities of the environment. Surroundings of course impose certain insurmountable limits within every cultural complex, and perhaps it can be understood that something like an absolute limit exists which can never be exceeded, but for long periods in the history of peoples this limit has been sufficiently far away for there to be no point in mentioning it in any equation. Other limitations imposed by low energy averages or technical insufficiency could, by contrast, be rapidly got around by some new discoveries capable of abruptly modifying the total of the K factor. And that means that the human species can modify the conditions of the environment which surrounds it in the sense of the tendency towards K factors growing in the long run. Would this mean that the same historical tendency is translated into a rise in the values which sociobiology supposes to be linked by determinism to K groups?

In a way the question could be answered affirmatively. For example, it may be thought that what we understand by cultural progress also contains, in general, lines of evolution of moral values, however moderate and humble they may appear in comparison with technological progress. It is precisely the existence of moral progress which creates an analytical problem of importance: the compatibility between some values which are assumed by definition to be universal and eternal, and their empirically comparable fluidity and expiry. We will leave it for another, more suitable, moment. Now let us suppose that the expression 'moral progress' has no basic contradiction and try to investigate the relation which progress of this sort can have to a modification of K factors within groups. Although we accept the general thesis that moral values progress, it remains to be seen whether they do so in the way sociobiology suggests.

The history of the evolution of certain evaluative concepts such as *agathos* in classical Greece seems to confirm the sociobiological thesis about r and K groups. Homeric society was a marvellous example of the domination of competitive values in connection with an expansionist policy, linked of necessity to arms. Anyone possessing the virtues which characterize a noble being and guarantee success in combat is, naturally, morally acceptable. As Greek culture obtains increasing resources, that is, as its K factor grows, the values of the *polis* substitute the agonal ethics of the ancient era for another in which *sophrosyne* and *hybris* are contradictory conditions which delimit the process of growing esteem for mutually binding virtues and communal respect. The term of excellence will end up by profoundly changing its meaning.[6] In the long run it was precisely the conflict between the exterior expansionist policy of the Athenian empire and the values which define the moral space of the *polis* which ruined the ethical sense of Classical Greece, and which, besides, contributed to undoing the hegemony of that type of society.

But in history we can also find equally useful counter examples. It was Wittfogel who destroyed naïve belief in the tolerance and benevolence of despotic oriental regimes. Some political systems which characterized the 'hydraulic empires' and formed those states capable of managing spectacular rises in the K factor of their populations from initial resources in an environment which represented very limited capacities, thanks to the development of technologies which still surprise us today. Between the political and state system and technological development there is a relation of necessity: only through despotism maintained by soldiers, state officials, and priests could the control and centralization of resources necessary to undertake works channelling and retaining of water be achieved, works which were essential to transform the conditions in which agricultural work was done in such a way as to initiate the system of growth of the K factor. Through a process like this new quantities of population are added to those already controlled, in a kind of self-sufficient growth, until once more a level close to the maximum population permitted by the environment now modified by hydraulic engineering is reached. The result does not however appear to be one of those idyllic villages with benevolent values. For Wittfogel the myth of benevolence serves a double function: it portrays the government and officials as pledged to getting the best for their subjects, which justifies the use of extreme disciplinary measures that are shown to be essential to get those optimum results. In addition it identifies the ethical ideal with the existence of the despotic state. Any local disaster, any type of corruption which comes to light, only indicates the unworthiness of certain officials and in the final extreme of the dynastic leaders. The excellence of the structural system of domination cannot be doubted. Thus, actions against the reliable elements of the state structure and even the very literary utopias which they construct (Wittfogel quotes the heroic myth of the *Shui-hu Ch'an*) only serve to support improved versions of the despotic state (Wittfogel, 1963, p. 163).[7]

The central body of values prevailing in that type of group which has carried its K factor to the limit can easily be summarized: obedience is the prime virtue, to the extent that in Mesopotamia 'good life' was identified with 'obedient life' and in Egypt survival itself depended on submission: 'opposition to a superior is a bad thing: while one is humble, one lives'. There is a whole ceremony which ritualises the gesture of submission, and which ethologists joyfully interpret as inhibiting norms of aggressive conduct capable of supporting the existence of hierarchies in the group. All that clearly entails fulfilling at least one of the conditions which we associated earlier with K groups: a long process of education in which the discipline of obedience begins to manifest itself within the family ranks. But to see evidence of worth accorded to benevolence in the cult of old age which could be developed would be a misinterpretation. Family obedience is nothing more than an element connected to the true submission in the group – that due to the official (Wittfogel, op. cit. p. 178).

The thesis that the evolution of evaluative content in hydraulic empires implies a phenomenon parallel to that represented by the establishment of the classical Greek *polis* does not seem to hold up – far from it. Rather it is the reverse, centripetal strengthening is accompanied by reinforcement of values quite distant from those of solidarity, appreciation of old age, and peaceful resolution of conflicts. Empirical evidence of growth in factor K with differential ethical progress obliges us to think of how the initial model of the continuum proposed by sociobiology must be transformed. Everything indicates that the human species counts on *two* possible strategies for the evolution of cultural norms referring to the r and K factors, which destroys the initial idea of a rigid environmental determination of moral phenomena situated on the gamma level. But, do we have to pause over this dual alternative or can we go on to detect different evolutions capable of reducing biological determinism to nothing and substituting it for one of a social type? I imagine we would soon reach the conclusion that we have placed ourselves in a line of argument similar to that posed many years ago by ethnolinguistics, and that it is not a case of trying to make an empirical list of existing moral codes in every society we might be able to study. The question in my opinion is framed in another way. If we must accept that norms and values contained in empirical codes are affected by determinants of a genetic or a social type, to what extent can the idea of the autonomy of moral phenomena which appeared to us to be linked in beta-moral terms to the critical character of language be defended?

Once again a radical alternative is raised here between an absolutely determinist postulate capable of leading to anthropological relativism and another completely autonomous one like the Kantian. Let us begin with the latter. If beta-moral autonomy is absolute and is not affected by social determinants of any sort, it can be assumed that the task of preference is foreign to elements coming from the social environment. These can be held as data to be taken into account when defining the setting in which certain values are opted for, but they will not influence the choice itself. The autonomic model expressed in this way is somewhat useless in explaining how proposals about moral principles are made; a historical example can make this obvious.

In the second of the *Treatises of Government*, in 1698, John Locke sets about the task of defining private property as one of the most firm and natural bases of human society. Locke, as is well known, is held to be a convinced contractualist. However, it is worthwhile not losing sight of how he takes the step towards contractualism from the existence of phenomena in nature. The natural world contains hard-working, virtuous people, but also criminals and robbers, and the law based on authority serves to impose order where crime tramples on rights obtained by nature. That is the meaning of rational submission to civil authority to which, through contract, comes a legitimacy derived from human, and in the final instance, divine nature. Moral defence of property and even its very origin are things which greatly precede civil contracts. God gave the Earth to all men, which means that they have a natural right to subsistence which reason, also natural, is able to disclose without difficulty. Certainly for Locke this does not mean that biblical references to lilies of the field have to be followed to the letter: the primitive right to the resources of the earth undergoes a correction through work and genuine natural law is that which grants the individual the right to the fruits of his labour, a law which underlines and gives force to positive laws which can be drawn up in this sense. What Locke is trying to state could not be clearer in principle. However, there are reasonable doubts about the significance of this law. Under the general concept of property, Locke understands at least two things: man's ownership of his own person and ownership of the product of his work done in nature. So the problem confronting us is that a citizen called John Locke is formulating a universal moral principle in the context of the civil English society of the Orange monarchs after the revolution of 1688, and if we do not take that into account and limit ourselves to the formal analysis of his theses we could verge on the limits of the absurd.

In his discourse, Citizen Locke includes some questions which from our current perspective on the universality of the norm would be extremely suspicious. When for example he enumerates the cases that lead to private property, he says "Thus the Grass my Horse has bit; the Turfs my Servant has cut; and the Ore I have digg'd in any place where I have a right to them in common with others, become my *Property*, without the assignation or consent of any body."[8]

To suppose that Citizen Locke was mistaken in likening his serf to a horse and distinguishing between the ownership of personal work according to whether it is a serf's or a gentleman's is naïve. Locke was not all mistaken; his standard of morality is in no way foreign to the context which surrounded him. To validate the moral judgement underlying this definition of property it is necessary to add a consideration of social determinants; the same as must be adduced for example for a reading of *Nicomachean Ethics* not to convert Aristotle into a merciless monster unable to understand that his theory of friendship is incompatible with the existence of a slave society in Athens.

A similar question could be asked with regard to the supposed tautology of certain universal norms. Taking an example from Paul Weiss, Maria Ossowska shows that the norm 'I will not kill my friend' is tautologous and not moral since the definition of friend necessarily includes that prohibition (1971, p. 115). If friendship itself contains things such as refraining from causing harm, it is tautologous to base such a norm on it. However, objections could be raised. The concepts of *friend* and even *brother* could in fact have an ethnic or simply social sense, in such a way that the condition of friendship or consanguinity is automatically established without the need for intimate contact and through the existence of *ad hoc* signs. The totemic system for example acts in this way.

We have another clear case of conceptual metaphor in Hebrew ethics. Despite the dispersion of the Greek term *ethnos* in different historical and religious texts of the Jewish people, that word takes on, above all in Philo of Alexandria, a sense which does not have specifically political or territorial connotations since it includes the colonies of the Diaspora. What defines *ethnos* is a 'supreme kinship' which as Philo said consists of one law, a single citizenry and the same God for all Jews (P. Geoltrain and F. Schmidt, 1978). As we know, this sentiment of *ethnos* could provoke serious problems of incompatibility with the civil power of the *pax romana*. So, Hebrew axiology expressed through the systems of aptitude and purity (as appears for example in Clévenot, 1976, p. 4 and 1.5) can lead us to certain confusions if we try to ignore the sense of *ethnos*. The ethical system of aptitude/debt includes assassination and slander among the ritual prohibitions. And these are prescriptions which come from the existence of closed clans in the North of Palestine, but they have been moved through the system of *ethnos* to a community in which the sense of kinship has varied in such a way that it cannot retain the slightest biological sense. 'I will not kill my brother' here takes on a sense far removed from ethnic groups and so is not very vulnerable to Weiss' objections.

In the previous chapter I criticized two theories designed to interpret preference (Rawls) and the argumentative character of ethical judgement (Perelman), precisely because they do not pay sufficient attention to the presence of this type of determinant which without doubt is capable of affecting the action of moral choice to a not insignificant degree. If determinist theses are accepted, however slightly, are we not converting rational preference into a utopia and decanting analysis of moral phenomena through the channel of relativism? In my opinion there is no need whatsoever to draw these conclusions, providing two things are made clear: a) what the relation between different moral levels is and what type of intrusion can lead to each of them on the level which has the possibility of autonomy and b) how can ethical preference be endowed with a base firm enough to convert moral progress into something separate from the determinist mechanism. This last question will be dealt with in chapter 8 once the last of the proposed levels, the delta-moral, has been examined. Let us now go on with the first.

In the first chapter of this book we said that moral conduct was too complex a phenomenon to be analysed as something homogeneous, and that in consequence it was necessary to distinguish between four levels of the moral phenomenon, which since then we have established. But an action like that of moral choice also undoubtedly has a sense as a unit about which certain judgements can be made. In fact it is this final result which is significant in empirical terms each time a doubt can be raised about the morally acceptable or rejectable character of things such as abortion, euthanasia, the policy of income distribution or apartheid. Acceptance or rejection qualify moral conduct in its entirety, but that does not mean that ethical judgement has a single harmonious interpretation. On the contrary, empirical moral preference is the result of factors which come from all the levels analysed, and as that is so, it is linked in a complex way to things which in no case must be assumed to be homogeneous or harmonious in their turn.

The multiplicity of factors come from the four levels (or, as we shall see, from structures which can be reduced to three of them), but need not be confused with the levels themselves. Motivation for example is a phenomenon which typically belongs, as we have seen, to the alpha-moral level and in analysing it we established the existence of biological determinants to a high degree which are also responsible for the human species having ethical conduct. But in no way, as I imagine I have already made clear, can it be assumed that empirical moral conduct is in itself also subject to such determinants. If that were so, all the arguments contained in the beta-moral level would be superfluous and, excepting different ecological circumstances, we could only talk of different gamma-moral level norms with difficulty. But the empirical moral presence is the result of a combination of factors which belong both to the person's character and the values present in the social group to which he belongs (with the addition of choice itself which is founded in beta-moral autonomy and obviously can be something of great significance or merely a trivial addition according to the case in question). That makes ethical choice a confusing box of odds and ends in which things difficult to reconcile are mixed, but in fact this is the character which we can empirically assign to such a conflictive phenomenon. It may be that the tendency to accept socially established ethical norms and follow typical patterns of behaviour make moral choice into something exceptionally rare in everyday conduct, but we cannot forget two things: a) that at any moment the actor can consider the moral meaning of his actions, since he has the elements with which to do so, however much they may be subject to his personal decisions, and b) that moral responsibility for the criteria of observers exists, despite the fact that the usual mode of conduct possibly has not posed the meaning of that responsibility. In human codes exemption from routine does not exist, or at least there is no such thing to the extent that all custom and all tradition are susceptible at least to internal criticism, something which I suppose even the most radical relativists would accept. One cannot escape the outcome of a car accident by referring to the habit of going at high speeds with little care, because although driving has become ever more routine, with greater velocity, the key element of responsibility for the act remains.

This moral choice, which is founded in such a complex way on multiple elements, acquires, as I said, a final result as a unit. Are we to suppose that this is the final balance of a kind of conjunction of factors, in such a way that all levels of the moral phenomenon end up by guiding the actor in a homogeneous and coincidental way towards the final conduct? The question would be answered affirmatively by all those who, with Darwin, insist on conceding great importance to the biological content of the alpha-moral level and conceive ethical progress as a confluence in which nature and rationality work together in harmony. Such a train of thought cannot be supported from the panorama set out here, which obliges the existence of conflicts between the different levels to be posed not only as something possible, but also as something which is relatively easy and extremely widespread. Let us summarize the mode of behaviour which an actor affected by elements coming from each one of the levels would follow if they were taken separately:

– *Alpha-moral level*: ethical conduct highly controlled by biological determinants which shape and confer the 'character of the individual' in its innate aspects. There are other factors determining alpha-moral conduct, above all environmental ones introduced in the process of endoculturation, but since the acceptance or not of indoctrinations (and the level of their acceptance) is in itself an innate feature, it is necessary to concede to genetic determinants a considerable part of motivation coming from this level.

– *Beta-moral level*: ethical conduct which is only slightly controlled by biological determinants. These only affect the conceptual structure which is of course indebted to those innate bases of moral language. Sociological determinism is higher, since we must attribute to it the subtlety and flexibility of ethical reasoning which from a historical and sociological point of view has been established, under threat of otherwise falling into gross errors of perspective such as those coming from the conceptual idealization ridiculed by Collingwood. But since the beta-moral level contains the elements of autonomy of judgement, those determinants cannot be so great as to make ethical progress impossible as long as we consider this level in isolation.

– *Gamma-moral level*: ethical conduct which can vary between very controlled and scarcely controlled (in comparison with previous levels) by social determinants. The circumstances of the group can favour the presence of open or rigidly closed moral codes, something which through the environment in which the individual is acting undoubtedly affects the possibility of following the socially instituted norms either closely or through a critical approach. If the theory already put forward about the gamma level is consistent, biological determinants in contrast would be low and limited in essence to the role played by the genetic code in providing norms with hierarchies and mechanisms of indoctrination for individuals of our species. This is the same as accepting a model of the 'Promethean gene' (in Lumsden and Wilson's terminology) for human cultures.

– *Delta-moral level*: considerations identical to those of the previous level.

In such circumstances (and if we reduce the last two levels to a single one for purposes of the analysis of preference) we can formally postulate the presence of three types of conflict in which two levels oppose each other, for reasons which will become clear in the next chapter. Moral preference determined by the alpha level through biological impositions can enter into contradiction with that which is controlled by the beta-level, and which adheres to lines of rational judgement, or with that which from the gamma level follows social norms, and thus an individual terrified by the presence of an injured person on the road can hesitate between taking heed of his wishes which impel him to flee, or stay and help the victims. In turn the beta and gamma levels can also enter into conflict: the driver can attend utilitarian considerations and abandon the injured, or on the contrary can defy prevailing norms of conduct by assisting, at great risk, victims who are considered racially inferior and contemptible in the socially established code. In fact, the phenomenon of moral progress can be interpreted as a conflict between the beta and gamma-moral levels from the individual perspective. But those formal possibilities also lead to empirical conflicts: the examples shown are not taken from science fiction. As this is so, human moral choice is affected by such degrees of ambiguity and doubt that it is not strange to find that throughout the history of ethics so many and such radically different systems of interpreting this complex phenomenon have been proposed. For however little one wants to make a stand on the importance of one of the levels over the other, hedonisms, deontologies and utilitarian rationalisms – or simply attempts to reduce ethics to sociobiological parameters – will appear. All these systems in isolation are able to explain moral phenomena from the perspective of one of the established levels, but while nothing resembling a harmonious conjunction of the different levels exists, in all formulations which fail to single

out that fact with sufficient precision serious lacunae will appear.
I am not claiming that the whole history of ethics is to be abandoned as useless and false – something which would probably be
resolved through suspecting my own theory to be more useless
and fallacious. The distinctions between motive and criterion, for
example, appeared precisely in order to overcome different aspects of the moral phenomenon and they represented in their day
a notable advance over the ethical systems which did not take
them into account. But I fear that today, with our increased
understanding of nature and human society, that is not enough to
be able to offer an effective interpretation of phenomena as
complex as moral ones.

By introducing conflict between the different levels we can
explain some of the paradoxes aired by sociobiology. Let us
return to the question of altruism. When in the first chapter
reference was made to the possible meaning of altruistic conduct
in order to base upon it a strategy for the study of moral phenomena, the need to admit (for reasons of evolutionary proximity) the
presence in human behaviour of a certain measure of genetically
determined altruism similar to that which might be present in
other superior gregarious mammals was raised. It was a case of an
act purely justifiable in alpha-moral terms, and thus it would be
necessary to consider for example automatic, or almost automatic
reactions such as jumping to the rescue of a small child from the
wheels of a car or the water of a swimming pool. Aside from
whether one accepts Trivers' or Dawkins' explanation to justify
this act from the genetic point of view, both the theory of group
selection and that of the selfish gene could give an account, of
greater or lesser weight, of all instinctive actions which as such
are genetically determined. Let us recall that Konrad Lorenz
demanded a biological imperative' of a type which could explain
the survival of hominids who wielded choppers from his perspective of the sharp distinction between aggressiveness and cultural
weapons.

Alpha-moral altruism can find itself in serious difficulties because of the existence of the other levels. With enough sang-froid
a father armed with an accurate high-powered rifle could save the
life of his son about to be trampled and perhaps gored by a bull.
But would an individual who faithfully followed religious pre-

cepts in which the animal who threatens the life of his son is considered sacred and inviolate choose a similar course? Of course there is no standard answer. One would have to consider many factors, among which would be things like the possible relaxation of traditional values because of cultural expansion able, among other things, to supply sophisticated hunting weapons. But that is the least of the matter; I am not interested in knowing the final behaviour, but in indicating the presence of conflicts between the different levels. A very radical case would be that presented by Alexander in the model of moral learning as a path towards utilitarian deception (cf. Chapter 1). The gamma values accepted and used in our societies,[9] if we are to heed Alexander, (and for my part I see little reason not to do so) would mean a guide for preference which keeps itself going on high measures of a calculated and refined egoism of a situational disposition. With its help, the individual is able to finally distinguish between useful types of behaviour and ones less useful for his interests, and in that way he would learn to deceive and hold himself back in his personal interest. If we must concede that the model exaggerates a little, no matter. The fact that this system of forming criteria of behaviour is sometimes used is enough to have to admit that, although it is not typical of all gamma values detectable in our moral codes, it is certainly one of the possibilities and can furthermore be empirically traced with certain ease. If the final conclusion is that in reality gamma altruism is a fiasco and restricted to hiding a refined egoism, no matter: what is important is not the name we give such behaviour, but the confirmation of the existence, under normal conditions, of severe contradictions between that almost suicidal alpha-moral altruism and these gamma-moral values.

Accepting the conflict of levels does not mean prophetically recognizing moral relativism. It means simply that empirical conduct is going to be determined by biological and social circumstance capable of perhaps imposing certain actions. Evaluation of such conduct will also be affected by the 'historical audience' and so we must once again insist on the need to have causal approximations which would allow us to escape the vicious circle of out-and-out formalism. That however does not lead to relativist solutions. Relativism demands a gamma-moral imposition on beta conceptual practice, and I think I have already argued sufficiently in favour of beta-moral autonomy. Such autonomy implies that moral progress is conceptually possible and it gives the foundations to justify it (remember that this is still pending) but does not convert it into something universal and inevitable. No open theory of ethics can afford to make the mistake of forgetting history.

CHAPTER 7

THE DELTA-MORAL LEVEL: GODS
AND GENES

What, we are then compelled to ask, made the hypothalamus and limbic system? They evolved by natural selection. That simple biological statement must be pursued to explain ethics and ethical philosophers, if not epistemology and epistemologists, at all depths. Self-existence, or the suicide that terminates it, is not the central question of philosophy. The hypothalamic-limbic complex automatically denies such logical reduction by countering it with feelings of guilt and altruism. In this one way the philosopher's own emotional control centers are wiser than his solipist consciousness, "knowing" that in evolutionary time the individual organism counts for almost nothing. In a Darwinist sense the organism does not live for itself; it reproduces genes, and it serves as their temporary carrier. (. . .)

Samuel Butler's famous aphorism, that the chicken is only an egg's way of making another egg, has been modernized: the organism is only DNA's way of making more DNA. More to the point, the hypothalamus and limbic system are engineered to perpetuate DNA.[1]

E. O. Wilson, *Sociobiology* (1975)

Ultimate goals are present in the cultural heritage of a human group, as are all other positive norms. Just as it is obvious that not all groups have identical normative codes, it is clear that they do not all invoke the same supreme values. Even so, there is reason for granting separate treatment to ultimate ends and reserving a complete level of the moral phenomenon for them. That reason has to do with the manner in which moral language is used. R. M.

Hare has placed special stress on how moral judgement is necessarily linked to its possibilities of universalization, and how that is achieved through an implicit or explicit task of reasoning (1963, pp. 86 ff.). To maintain a moral judgement implies that the person who makes it can offer reasons for defending it. If someone holds that abortion is an ethically reprehensible act, he might be required to explain why, and thus begin a chain of reasons which can lead, for example, to maintaining that, in this way, the life of a person is terminated and as such it is always a condemnable act. Perhaps the normative discussion might lead to doubting whether a foetus can be regarded as a 'person', with which we would digress from the main line of reasoning, but the interlocutor might also raise arguments about how, in certain cases, killing might be morally acceptable, or call upon utilitarian reasons about the life of the mother for example, or the impossibility of the family adequately feeding and bringing up more children, or other arguments along these lines. The discussion would probably reach a stage in which those arguments would meet a final obstacle if the refusal to accept abortion makes reference to a supreme and indisputable value such as the will of God or the consideration above all other circumstances that the life of a foetus is an end in itself which cannot be subordinated to any other. To call upon the ultimate end means to get around the possibilities of following ethical reasoning, and that circumstance demands special attention.

But, can one really talk of an ultimate end? What does it mean? The history of ethics, if we understand ethics to be a systematized philosophical discipline, begins precisely by asking, in the opening lines of the *Nicomachean Ethics*, what might be the good towards which all things tend and by taking for granted that an ordered activity of reason will be able, at least, to give some lead in this respect. Initial optimism has not always been supported. Existentialist anthropology, for example, insists on proclaiming the 'end of ethics' precisely because it claims that the Aristotelian quest, be it directed towards God, man, or nature, is a folly, and that each person would do well to organize his life with as much dignity as possible, but without greater pretensions. Between both extremes there are innumerable proposals about what this end towards which everything tends might be, and one

of the most novel is that stated by sociobiology, slightly transforming the classic Darwinian postulate of 'the conservation of
the species': what we do or do not do, is, in the final analysis,
designed for the task of perpetuating a double helix with four
nitrogenous bases.

All these ultimate ends have one common characteristic. They
claim to be absolute, that is, they do not depend on historically
variable circumstances. However, that is not a character which
can be conceded without further discussion, given that some of
the proposed ultimate ends are incompatible from a logical point
of view with others which leads us to the necessity of accepting
them all in the name of ethical relativism but denying them an
absolute character, or alternatively to the thankless task of finding the means by which the 'true ultimate end' might be recog-
nised. Needless to say, a rationalist theory of ethics should be
able to provide adequate means for that purpose. There is no
shortage of attempts to achieve this, such as that of Peter Singer,
who with considerable acumen criticized Wilson's proposal for a
radical biologization of ethics, warning against the trap into which
sociobiology falls. If ethical judgement is nothing more than the
result of the activity of centres of emotional control it is inevitable
that one should fall into out-and-out relativism: there would be as
much as reason to criticize ethical preferences as gastronomic
preferences (Singer, 1931, p. 85). Singer proposes a rationally
based ethics as a counterbalance, a proposition which has all my
sympathy, and he puts forward the possibility of using reason
against supposedly absolute postulates (as biologists do) invoking
the differences between the person immersed in a specific cultural
tradition who has a 'participant's perspective' and the one who is
given to more objective judgements from an 'observer's point of
view'. This mechanism does not seem able to dispel all doubts,
above all because the same participant/observer alternative has
been used with exactly the opposite intention of converting rationality into a useless mirage.

In *Science in a Free Society* (1978, p. 20) Feyerabend states that
rationalism is a secularized form of belief in the power of God's
word. Given that it would be easy to find authors who are not
prepared to grant rationalist positions even the benefit of secu-

larization (if it transpires that what is 'lay' is of benefit in some way), we who consider ourselves to be at least a little rationalist should be grateful to Feyerabend. But perhaps not so grateful that we accept his thesis too readily. The sense of rationality in Feyerabend and the relations which could exist between reason and practice run along a path of comparison between traditions which is derived from two types of questions: questions by the observer and questions by the participant. Observers ask *from outside* things such as what is happening and what is going to happen; participants ask themselves *from within* whether they really have to take seriously this discussion which seems so idiotic.

Since this dual framing of speculative attitudes towards any subject present in a group has already been supported, normally by anthropologists under what they call the *emic* perspective (that of the participant) and the *etic* (of the observer),[2] and taking for granted that there is no reason to claim that what Feyerabend proposes is essentially different, we can carry on using the traditional terminology. So, Feyerabend makes it clear that what we call reason and what we understand as practice are in reality two different types of practice. It is the habit of adopting the etic perspective in the activity of reason, and the emic in practice, which leads, according to Feyerabend, to the final result of having two artificially separated factors. As opposed to such a strategy, by any reckoning reprehensible, Feyerabend defends an interactive conception of practice and reason, which links ten theses aimed at explaining what the factors of this interactive process are like. I will refer to only two of them, to which the others are undoubtedly related:

I. Traditions are neither good nor bad, they simply are.
II. A tradition acquires desirable or undesirable characteristics only when it is compared to another tradition. (1978, page 27).

When Feyerabend explains the first thesis he seems to be indicating that he refers to tradition from an etic perspective, hence if we were to use the emic it would be normal, within reasonable limits, for the average participant to choose to rate his own tradition as 'acceptably good' according to the rules for assessing good and evil which all traditions have. If it were not so in general the tradition would not last long and would not interest us. But surprisingly the corollary Feyerabend draws (from his arguments, not mine) is that rationality is not an arbitrator of traditions but one more tradition, or even only an aspect of tradition. It is neither good nor bad, it simply is. But that existence is of little use, at least in terms of intervening in the comparison between traditions. If a comparison is finally made, and moreover with claims of rationality, it is through the fault of the second of the theses: rationalist participants, befuddled beneath their emic vision of the world, dare with etic pretensions to criticize other traditions.

One might expect Feyerabend to realize that his own formulations are irremediably affected by emic aspects. However, it is not obvious that he does. For Feyerabend, we have not reduced possible activities to the only final emic sense surviving the second thesis; nor even have we exiled every trace of objectivity. The relativism of Protagorous which Feyerabend supposes can be deduced from theses I and II is *reasonable* because it takes account of the pluralism of traditions, and *civilized* in that it does not suppose local customs to be by definition those of the greatest excellence (1978, p. 28, third thesis). It seems that a prudent dose of relativism in this respect is enough to lead to a special form of emic perspective or attitude which would be that of 'pragmatic philosophy'. A pragmatist is a sort of participant who, in addition, is an observer who realises the ephemeral nature of traditions and norms and so does not give them absolute value. A pragmatist is, in short, a very difficult individual to find.

I wonder if we have found one in Feyerabend himself. His perspective is, as I said earlier, emic, at least in one sense. But his critical vision of the useless and mistaken world of a rationalism attached to its traditions makes him think that attachment in general is a sterilizing attitude, and wield, in consequence a liberating etic aspect. Let us imagine that this is even possible (something of course rather risky from Feyerabend's own perspective and in view of his two theses) and that rationalists are completely mistaken in their original theses. So, from what to me seems to be included in what we could call the ideal strategy of rationalism, that would be enough for those who brandish reason as a flag to convert urgently to Feyerabendism. There is nothing more rational than to accept that which under logical arguments is seen to be valid and convincing.

But, is that all? If we identify the rationalist cancer with the philosophic doctrine developed under that name in the modern era, or with the alternatives to Homeric common sense which appeared in Greece between the 6th and 4th centuries B.C., or even with the contemporary theory of science which talks of great criteria, the matter would be easily remedied by conversion. Let us forget Descartes, Plato, and Popper; if necessary let us leave aside Spinoza, Aristotle, and Lakatos. Are we then free of rationalist difficulties? I am afraid not. While we set about proclaiming that traditions are good or bad according to a criterion which we still suppose to be only partly objective (maintaining for example that to kill Jews or Palestinians soley because they are Jews or Palestinians is a mistaken and reprobate tradition); as long as we think that our own tradition contains desirable or undesirable properties aside from any type of comparison (postulating, if the case arises, that an arms policy is an incitement to collective suicide), we return to the vices of rationalism.

However, these so-called vices are the only ones which can contain the point of support necessary to escape the cul-de-sac in which ethical relativism leaves us. If we accept that there is nothing resembling a mechanism capable of guiding the comparison between traditions in an etic sense and weighing up the respective proposed ultimate ends, the discussion has ended. The most we can do is invoke the principle of objectivity and go ahead with the task of criticizing the rationality of the means, since the ends in themselves have remained confined to the security of the limbo of the just. But before throwing in the towel, I should like to go a little further into the relations between rational attitude and emic and etic perspectives. I understand that under the classification of 'rational individual' we can put all those who, faced with a given stimulus from the environment, act in order to satisfy their aims (aims understood in an internal and direct sense, irrespective of whether through analytic activity we could claim that their 'real' aims are different) in accordance with the conduct which the tradition in which they are immersed shows to be more

favourably adjusted to the attainment of those aims.[3] The significance of what I am trying to say will be better understood with an
example. Evans-Pritchard tells us in *Witchcraft, Oracles and
Magic among the Azande* (1937) that the Azande do not believe
that misfortune occurs by chance. When someone is ill, or dies, it
is due to witchcraft (Chapter 1). The Zande who is *pani nagbera*
(in a bad condition) has ways of knowing who among his neighbours is casting a spell on him: he consults exorcists or goes to
oracles, which are cheap and suspect (like the oracle of the
rubbing board), or expensive and more reliable (the oracle of the
poison). The Zande is superstitious, but not ingenuous. He will
not dare seek revenge on the basis of unreliable heuristic methods
such as those used by exorcists, and in an extreme case will use
the onerous and complicated system of poisoning chickens to
know if a specific person is responsible for the witchcraft.

What is rational behaviour in this case? Rational for all Zandes
is to suspect witchcraft in the case of misfortune, to consult
oracles and exorcists to narrow the field of suspects, undertake
the oracle of poison in one or two cases and then, as a result, seek
revenge.

Irrational behaviour would be just to send a chicken wing to a
presumed sorcerer to accuse him on the evidence gained from
exorcists alone. Or to throw oneself headlong into the sacrifice of
fowl without having the slightest idea where the blame might lie.
This the Azande do not do. Their traditions show them clearly
that this is stupid behaviour and not only leads to expense, but
worse, allows witchcraft to act with impunity. The world of the
Azande is a sea of suspicion and revenge, but from the emic point
of view that is the only rational possibility.

And from the etic? An anthropologist will probably opt to take
quinine in the case of malaria rather than undertake fatiguing
journeys in search of the climbing *benge* plant with which to fill a
chicken's crop. Certainly quinine implies another, different, tradition, with which the anthropologist's gesture can in turn be
considered as emic, but as regards comparing results of one
activity with another, the perspective is clearly etic. The anthropologist would not stick firmly to his quinine if after many failures
a Zande friend of his solved the problem by oracles and revenge.

From an etic perspective two very different things can be said about the behaviour of the Azande:

1. It is irrational, because however much you poison chickens and search for witches there is no cause-effect relationship with the real problem.
2. It is rational because it corresponds to what any person in the position of a Zande would do.

With a view to simplifying the panorama I will henceforth call etic rationality that which demands judgements like the first, and emic that which conforms to the second.[4]

So, the thesis I will defend is the following: irrespective of whether one can formulate reasons from an etic perspective, emic rationality has a genetic function. It is determined by the way in which natural selection acts in the human species, and entails a setting of genetic-social determinants which imply, in a broad sense, a tradition. Tradition in turn remains subject to selective laws.

Before going into the consequences of this thesis fully, why a man acts through emic rationality will have to be explained. Given the possibility of certain behaviours, could he not understand which is the most rational – aside from whether he formulates the question in that way of course – and then act in another way? Of course he could. If not cultural progress would be impossible or in any case extremely slow, because it would only be justified through mutations in the genetic determination of rational emic conduct. What I maintain is that man is, to a certain degree, conditioned to prefer what in his tradition is assumed to be rational conduct, in the sense we are using the term, and that preference works as a more or less strong tendency by virtue of mechanisms (usually taboos) which limit other possible conducts. An individual may act in a different manner from the traditional because he is playing, because he is mistaken in his interpretation of stimuli from the environment, because he interprets his tradition badly, or simply because he wants to. But overall, in a perspective of populations, emic rational conduct is standard. Why? For the same reason that eye-brain mechanisms do not

exactly reflect the outside world, but do so with an approximation
necessary for the form of individual/environment interaction
typical of the species' way of adaptation. The theory of evolution
forbids representative hallucinations on pain of extinction. And
the adaptive path of man passes through the need to assimilate
traditions and accept them as positive elements of meaningfully
useful behaviour. Critical conduct is thus moulded by tradition,
which lends the general sense of rationality, whether this is
accepted or not later by the individual in a given case. In that
way, man adapted himself to the environment and the tendency
to construct traditions is as significant from the genetic point of
view as the tendency to accept as valid what in general has
already been demonstrated to be so.

Let us turn now to the consequences of the theses expressed
previously. Those which refer to genetic determinants that can be
associated with rational conduct are what can most influence the
central question which now concerns us. In what sense is that
conduct genetically determined?

To answer this we have to use a very well known distinction
made by Horkheimer regarding the way what is rational acts in
the course of human choice. Final rationality refers to the formulation of ends or ultimate values, while the instrumental is restricted to the rational conduct necessary to achieve those ul
timate ends. Given that the empiricist tradition proscribed discussion of ultimate rationalities by virtue of the principle of objectivity, let us begin with instrumental reason.

We are not going to create problems for ourselves. Given a
specific end, rational conduct which serves as an instrument is
always linked to reasons which will solve the problem of discrep-
ancies. A Zande disposed to cure completely a persistent severe
headache might have a conversation such as the following with a
disciple of Evans-Pritchard (in the utopian case that the Zande
culture had not already been wiped out).

Anthropologist: "What are you doing?"
Zande: "I am making a brush with *bingba* grass for the poison"
A.: "Are you going to consult the oracle?"
Z.: "Yes"
A.: "Which formula do you use?"

Z.: "It's very simple. The *batamba singa* is the first test: 'If X has cast a spell on me, oracle of the poison, make the fowl die'. I administer the poison. If it dies I use the second test, *gingo*: 'The oracle of the poison has said that X is guilty; if the answer is correct, make this other bird live'. If the second one lives, X is the guilty party."

A.: "And if the second one dies?"

Z.: "The oracle is useless."

The anthropologist could discuss ritual aspects which might mean a loss of credibility in what the oracle determines, but in no way would it make sense for him to explain to the Zande that what he is doing is ridiculous and that what is in fact happening is that he has a malignant tumour. Even if he were to do so, the Zande would remain convinced that the guilt lies finally with X, something which must be proven. If the anthropologist makes use of the cause/effect mechanisms of modern interpretation of illness, he is moving away from instrumental rationality. This cannot go against the basic principles of traditions, against primordial beliefs understood in a sense somewhat different from that used by Avrum Stroll, but identical in so far as they form a certain way of interpreting the world. If witchcraft disappeared from the Zande horizon, the world of the Azande would not be the same.

Emic instrumental rationality contains, then, certain genetic determinants which refer not only to how man infers ways of interpreting the world and accepts and rejects changes in them, but also to the very existence of emic reason as a means.

But the question of ultimate ends cannot be got rid of so easily. The absolute claim of ends gives them an apparently etic character, so that an ultimate end which really was just that would not be unduly affected by any specific tradition. To postulate divine will or the duplication of DNA has this transcendent character. However, these two ends are not absolutely homologous. Divine will as an ultimate end eliminates rationality in a different way than genetic exigency. God as a supreme value demands an act of faith, a refusal to discuss the matter. Not in vain is Heidegger decisive in the introduction to *Was ist Metaphysik?* (1929), when he exhorts believers to take the words of the Apostle who relegated philosophy to the category of madness seriously for once.[5]

To propose DNA as an ultimate end however is the result of a task of rationalization of the mechanism of life from the biological point of view. Although for Wilson that would also mean leaving aside philosophy, transcended by a 'hidden intention', an identical end for all living beings which, thanks to some specific conditions of the human capacity to reason has been laboriously made clear.

But to locate the task of replication of DNA in the supreme delta-moral level can be an error of perspective about what an ultimate end from a philosophical point of view really means. In his commentary on Waddington's *Science and Ethics* (1914), D. D. Raphael denounces the tendency of biologists to confuse what is *going to* happen with what *ought to* happen.[6] Undoubtedly in statistical terms, and expecting catastrophes, human beings are going to transmit their respective combinations of DNA to the following generation, but in no way does this mean that the transmission of DNA is a moral duty for any of the specific beings who populate the earth. Perhaps there is a religion I am unaware of where the transmission of DNA is considered the supreme ethical value, but being so, it would be a very different question to that posed by Wilson.

Through the manner in which the species has evolved, that is, the way the specification contained in the chains of DNA which feed human genes has been modified, certain relations between genetic inheritance, language and conduct have been established which up till now have led us to establish the presence of genetic determinants in the alpha and beta-moral levels. By virtue of these relations, and the very character of the presence of moral language in the traditions maintained by human groups, certain values held to be absolute are proposed, and their diversity is obvious. Each ethical system proposes an ultimate desirable end, and if Wilson established his own ethical system, something which he does not seem to have done, perhaps it would contain a different one from other systems. Another biologist, probably less approved of by the scientific community, could postulate as an ultimate end that individuals should breathe as often as possible. So the *inevitable conclusion is that all delta-moral ultimate ends are emic ends*. If we must accept for now one genetic determination in this field, it is the way in which DNA has endowed human survival machines with a notable capacity to propose alternative ultimate ends.

But I am not at all in agreement with Feyerabend on the impossibility of making an internal criticism of traditions, or a comparison between them. To concede to ultimate ends a hidden emic character beneath their absolute etic claims does not necessarily mean falling into moral relativism. It means the need to offer some type of theory able to defend the preference between ethical codes, sufficiently strong to arrive at the criticism of ultimate ends (something which will have to be done immediately). I will try to show how the theory of preference between alternative codes allows one to understand that ultimate ends have to be considered emic because of the way in which moral language is organised, and can be compared and ranked, within certain conditions, thanks also to the semantic structure of moral terms like "Good".

Chapter 8

MORAL PROGRESS

> As society has always moved in class anta-
> gonism, morals have always been class
> morals, either they have justified the
> power and interests of the dominant class,
> or, since the oppressed class became strong
> enough to do so, they have represented
> revolt against that domination, and the
> interests of the future of the oppressed.
> There can be no doubt that overall moral
> progress has been produced, as in all
> branches of human knowledge. But we
> have still not overcome class morals. Truly
> human morals, above social class antago-
> nisms and its legacies, will only be possible
> in a society which has not only overcome, but
> in practice has actually forgotten conflict
> between social classes.
>
> F. Engels, *Anti-Duhring*

The way in which the criterion of what is right or wrong in ethical matters is justified in, shall we say, orhodox Marxist theory refers rather nebulously to the working class, and consequently has to bear accusations of fallacious naturalism. But Engel's text opens a new and interesting possibility, although of an eschatological nature, by alerting us to the advent of 'truly human' morals. That these morals should coincide with those of the working class, boosted by its political triumph, is something we can leave aside. The fact is that Engels accepts, even though he does not explain why, the existence of historical moral progress. Must we assume that the arrival of truly human morals will mean the end of such progress? Or, on the contrary, will growth in this sense continue to exist? And if so, under what conditions will we know that a given variation in ethical values signifies progress, if, by definition, the class antagonism which indirectly produced the criterion of choice no longer exists?

The problem of moral progress understood in a theoretical way, that is, as a formal possibility, coincides considerably with the situation which would appear in that idyllic society which Engels announces. Throughout this book reference has been made to circumstances relating to the ethical phenomenon which lead to the problem of moral progress, and the possibility of offering a formally adequate model to interpret it has been suggested. It is worth stressing the difference between a formal model of progress and its empirical existence. The conditions necessary to postulate progress can break down for reasons as varied as the multiple forms of subjugation and tyranny which human groups have managed to invent. All the impositions which Marx and Darwin analysed from social and genetic perspectives seem to indicate clearly that our present way of thinking out ethical problems shamelessly ignores the history of humanity. But, to the extent to which what is being discussed here is the formal model necessary to make ethical progress possible, taking for granted that this is also empirically feasible, we can clad ourselves in the armour of indifference. The essential problem lies in finding the way to understand what moral progress is, and how, in an ethical alternative, to detect which is the line of progress. When all is said and done, that is also significant for historical experience, since we must face up to the evidence that moral values do not remain unalterable throughout the centuries. If today the death penalty is considered unfeasible and some time ago it was acceptable, there has been a moral change. But has there also been *progress*, or do we have to resign ourselves to accepting relativism pure and simple?

As Ernest Tugendhat points out (1979), there is a paradox behind every historical experience of morals. Moral proposals claim to be absolute (in the sense in which we used 'absolute' earlier, see Chapter 1, Note 11) and at the same time experience shows us that individuals and societies manage, through empirical processes, to reach new moral convictions which are held to be more valid than previous ones, and of course once again as absolute. If I have not misunderstood Tugendhat, an ethical theory would have to either justify objectivity independently

from experience, or alternatively maintain that something we
could call 'historical progress' might exist, but it would at first
sight be paradoxical to do both, hence, just as the analytic
philosophy of moral language has insistently declared, there can
be nothing resembling an empirical foundation for morals. So,
Tugendhat's idea consist of ignoring the analytic prohibition
and determining the conditions which would have to be fulfilled
by an ethical system in order to confront both tasks: the claim to
being absolute, and historical progress.

I imagine it is not necessary to over-emphasize the similarity of
the task undertaken here. Between Tugendhat's project and what
we are trying to construct regarding rational preference there are
ties as evident as those which derive from the need to have certain
foundations of an empirical type, causally explicable, and the
need simultaneously to defend the character of ethical preference. Tugendhat considers that if a moral experience with absolute pretentions is feasible, it would be obvious from the integration of two factors: an *a priori* fundamental principle and an
experience analysable from that point of preference. The role of
one or the other seems clear in relation to what in Chapter 5 I
have called the 'asymptotic approximation' to a desirable end: the a
priori principle is able to establish, albeit crudely, the shape of
the curve to follow, while empirical experience attends to advis-
ing us if we are really on the road to progress. This idea can be
connected to the system of analysing morals which has been
defended up till now in the following ways: an 'empirical moral
experience' consists of the compatability of the beta-moral element with those of the gamma and delta-moral levels through a
process in which the two partial aspects of the beta-moral level
structure are made clear. Let us recall that these are maintained
through the following two theses:

1. The beta-moral level is biologically determined through the
   releasing function of evaluative words (primordially 'good').
2. Descriptive and argumentative functions add to the releasing
   character of moral language a second beta-moral aspect of
   indeterminism which supports an evolutionary strategy based
   on an 'open programme of conduct'.

The significance of the second of these theses is also limited. It

assumes the possibility that in the heart of the group empirical
elements of evaluation (gamma-moral elements) and even ultimate ends (delta-moral elements) are developed which may
imply disadvantages for the adaptive process. These drawbacks
do not of course mean so great a burden that the group has
insurmountable difficulties in surviving. I should like to stress the
non-tautologous aspects of the previous sentence, that is, the
empirical and historical possibilities that traditions and cultures
disappear, and part of the blame for this phenomenon must be
laid on the presence of 'unsuitable' moral norms. As I have tried
to explain, the possible presence of dysfunctional elements is
more than compensated for by the advantages of indeterminism
inherent in an open programme of conduct. That means, in
summary, that the beta-moral level is able to give the species
innate knowledge, an ordering of the world prior to experience,
which includes the use of evaluations through the concept of
goodness and that historical circumstances could fill the innate
base with different meanings for what, in each specific tradition,
is understood as being 'good' morally speaking.

Moral progress, 'empirical moral experience' is just a change in
the compatibility of ultimate ends (delta-moral) or the positive
proposals of the group (gamma-moral) with the global content of
beta-moral structures. In the details which now interest us, that
process is also beta-moral. In fact the aspects connected to the
second thesis are those which are significant. Without that critical
possibility, however little it might have empirical reality, there is
no sense in talking of indeterminants. Gamma and delta-moral
differences could be explained through things like population
genetics, of which we spoke in Chapter 6, but it is the activity of
adherence and preference which is decisive in producing moral
progress in the strict sense of the word. The innate foundation
supplies the *a priori* claimed by Tugendhat; argumentative indeterminism is able to complete the necessary. But what sense does
it make to say that this line of progress means a tendency towards
the growing compatibility of the two elements on the beta-moral
level? Simply, that the concept of 'good', able to direct primordial knowledge of the individual, is not empty. It contains what
we could call a dimensional sense, a channel along which the
different possibilities demanded by indeterminism are going to

run. In spite of what experience shows us through comparisons
coming from work in the field of cultural anthropology, there is
not absolute liberty to endow the concept of 'good' with content.
I should express myself better: a tradition is able to classify as
'good' almost any action, but under conditions which refer to the
form demanded by the structural significance of 'good'. Tugendhat shows that the semantic base of good is linked to the fact
that the proposal is good in general and not by virtue of private
criteria: 'good' obeys the impartial interest of all.[1] I have only to
add a small clarification: that semantic sense is beta-moral and
not gamma-moral. What it means is that within a group an ethical
norm with a gamma-moral content can be developed which in fact
serves the interest of a social class, group or individual. Beta
indeterminism permits it as long as an ideological system of
asymmetry exists within the group able to make the said norm
pass as impartial. If the individuals in the group accept the
structure of asymmetry (and that does not mean that they accept
it in reality because there is sufficient machinery of power to
impose the hierarchical situation by force, but that within the
group emic values exist favourable to asymmetry) and if, moreover, the individuals understand that the norm is suitable to the
structure of the hierarchy, the initial conditions are fulfilled.

Tugendhat is not so liberal in accepting ethical systems. For
him phenomena of moral progress are only meaningful when
empirical progress can in fact exist, that is, when the ethical
system includes non-authoritarian morals, rational morals. But
that in my judgement is to reduce matters to a single dimension of
the problem. In posing what happened on the gamma-moral level, I
referred to the existence of two strategies which both seemed to be
evolutionarily stable in human groups and in regard to ethical
content. Sociobiology relates them to r and k groups, but I
have expressed by doubts about the possibility of simplifying
matters so much. One strategy implies potentiating the adaptive
meaning of keeping up traditions and will then lead to authoritarian morals dominating. Another makes use of the advantages
of the critical character for adaptation, and aims more towards
'rational' morals, which are perhaps better classified under the
terms of open or tolerant. So, the matter of moral progress, that
is, of empirical experience able to modify normative rules, ap-

pears in both strategies – though under authoritarian morals a little more time might have to pass before it does. It would be rather difficult to propose the existence of societies which have completely overridden the aspect of moral indeterminism linked to the critical character of language, although literature has undoubtedly been able to create anti-utopias as beautiful as they are worrying, such as Orwell's *1984*.

Both strategies, the 'authoritarian' and the 'tolerant' *progress* in the sense of making awareness of fairness associated with what is termed 'good' grow. Thus conflicts relating, for example, to fiscal pressures applied by corrupt officials for their own benefit are resolved. Of course, that would only mean claiming that the categories proposed by Tugendhat are theoretical extremes, and that even in groups with authoritarian morality there is the possibility of a chink of rationality. In any case that is the least of the matter; the important thing is to be able to link the sense of progress to growth in the process of fairness and not to make utopian statements that this is necessarily going to happen in societies.

But let us forget questions relating to the gamma level and return to the formalization of a system of beta-moral preference worthy of credibility. If the panorama presented by Tugendhat and made possible, in my opinion, by crossing the two beta-moral elements is taken to be correct, the necessary foundations for the reconstruction of the now somewhat rusty and *démodé* mechanism of the rational preferer have been laid. There are etic means of rationalization of morals simply by resorting to the process of specification, as Tugendhat shows, a process which Hare links to strategies destined to universalize ethical proposals. An individual willing to assume rational ethical positions, if there are any, will be able to settle his conflicts by merely reducing the significance of norms in dispute, thereby increasing their universality. Let us imagine a case such as euthanasia. The starting point is the content which, for ethology, comes from the alpha level and implies the existence of a way of containing innate aggressiveness in a group through the genetically determined dictum (according to ethologists) 'do not kill a friend'. There are ways of identifying all those who should not be attacked, and of course some of those means have a sense which is analysable from the beta-moral level (all those not based on the use of universal signs of inhibition of aggressive conduct, like 'norms of greeting'). Our beta-moral schemes have led us, I think, to the conclusion that an acceptable gamma norm is that one should not kill anyone (do not forget that I am an optimist at heart). But, 'do not kill anyone' is a contradictory norm as regards my judgement that perhaps it is desirable to take the life of a friend of mine with an incurable illness who is suffering severe pain and asking me to inject morphine, and I know perfectly well that the injection will terminate his life. The norm becomes more specific and at the same time universalized if I accept 'do not kill anyone – except when he himself asks you to and he has no possible salvation', is preferable. I am not arguing about what is desirable or reprehensible in euthanasia, but about the way in which the norm becomes universalized. It will be necessary to consider of course whether whoever is requesting death is in his right mind, if the person giving the injection knows enough medicine to appreciate the consequences, etc. etc. And, is this not to fall once more into the trap of the infinite need for information which rational preference demands? Well, no, for the simple reason that we already know beforehand that the most one can hope for is an asymptotic approximation and that our judgement will never be the best of all possible ones. It is enough that it be rationally, or rather, reasonably, more adjusted to the impartial interest which underlies the structure of 'good'.

A question which could be posed finally is why should we have to imagine that the evaluative word 'good' has that type of structure. Tugendhat seeks the foundation in the word's semantics, but he links such a foundation to the presence of rational morals, and recognizes that this might, at bottom, be 'a mere pretension and an ideological illusion'. It seems to me that one could seek a less random causal foundation, as long as the semantic sense is disconnected from its link to a given type of normative justification and we look for its roots in another part. Biology can offer its support for this task through examination of the type of selection which is significant in the human species: an examination which makes us go back over a matter already mentioned in the first chapter of this book, that of group selection and inclusive fitness.

Man is one of the social species which has had the greatest

success in exploiting the advantages offered by the presence of the group as an instrument for achieving adaptation to extremely different ecological niches. We would be, then, an exemplar case of group selection as an alternative to individual selection, a concept (group selection) which arises from the difficulties Darwin had in explaining in what mysterious way evolution could select the existence of sterile beings in insect societies.[2] But the evidence for the advantages of group selection seems to be superior for the ignorant than for those who handle the question of survival in terms of population genetics. If anyone considers that hermites do not represent too valuable an example to explain to a Martian how man behaves in his life on earth, it still remains to be decided what the adaptive significance of living in a group could be. The first model which occurs to us is simple and seems final: with just a little comprehension and effort on the part of individuals, the advantages of living in a group are so great that they ensure the success of those who stay together. But things are not so clear from the genetic viewpoint. Let us imagine a group in which behaviour has been developed which we will connect, for ease of explanation, with an altruistic gene. The individuals in the group limit their own chances of survival a little in favour of the whole group. And, if a mutant individual loses this altruistic gene, and besides enjoying all the advantages of belonging to a group does not have to pay anything in return in terms of sacrificing his chances of survival? Well, simply the prospects of success for that mutant 'selfish' gene would increase so much that it would rapidly spread throughout the group. Anyone interested in the prospects of success for a gene of this type in a population has only to turn to Dawkins' synthesis in a highly successful book, *The Selfish Gene*. What interests me now is not the discussion of possibilities of altruism versus egoism in the gene-pool of a population, but rather what, since Wynne-Edwards, implies the maximization of the theory of group selection.[3] Wynne-Edwards maintains that through social conventions such as birth control, infanticide and hierarchical submission, groups progress at the expense of a little individual sacrifice which also causes individuals to progress in the final instance. It is the theory of the social contract transferred to population genetics: Wynne-Edwards, if we are to believe Wilson, considers that individuals do this *voluntarily*. For reasons connected to the strict application of the theory of natural selection – which operates through models of individual selection in the final instance – that is extremely doubtful. But there are certain conditions linked to population dispersion, to the existence of empty zones and isolation, where something resembling Wynne-Edwards' model could make sense: kin selection. If the members of a group are related to each other and this group forms, through isolation, a 'genetic population', the concept of *inclusive fitness* coined by Hamilton[4] alters matters. That inclusive fitness would be the sum of aptitudes which the individual possesses in order to survive, plus all the effects of his behaviour on his relatives. Given that the question of selection and adaptation has to be contemplated through the prism of the permanence of DNA, a useful strategy will be that which implies high amounts of inclusive fitness, and that can be best achieved perhaps under some reductions in individual fitness which can manage large increases in the effects of behaviour on the fitness of relatives. The difference between this model and Wynne-Edwards' is that now one's own genes benefit and not other, different genes which form part of meta-populations. Seen in this way, the gene which controls 'inclusive fitness' is, however much phenotype conduct disguises it, an extremely selfish gene.

How can the effect of that disguised altruism be measured? Through the relation coefficient which defines the quantity of genes shared by two people, statistically speaking of course. If we leave aside the presence of inbreeding, at least as a starting point, the chance of a given gene in an individual forming part of the genetic inheritance of his brother is 1/2. Thus, inclusive fitness has positive results as long as an extreme act of altruism, leading to the death of an individual, increases the possibilities of survival and procreation of his brother to the extent of doubling them, or preferably even adding a little more. That does not of course mean that heroes attend to their inclusive fitness, calculator in hand, to assign relation coefficients to any individual who might be affected by their heroic action before undertaking it. It means that the tendency of altruistic acts to appear, from the point of view of selection, is affected by the possibility a gene has of extending itself in the population pool. Only those genes which represent making use of inclusive fitness can prosper in the pool, that is, only those altruistic acts which are connected to the phenomenon of inclusive fitness can occur, in the long run, in the population. The model can get much more complicated because in groups related by kinship there is of course a situation of endogamy and the relation coefficient will not be so easy to calculate. In the same way, its expression will vary from 1/2 for brothers to 1/32 for a second cousin. No matter, that means that the calculation must simply be adjusted, which biologists find no difficulty in doing.[5] For our purposes it is enough to take account of the general sense of the theory of Hamilton: the altruistic act masks the progress of inclusive fitness tending to improve the possibility of survival of DNA.

The models of group selection and kin selection are alternative ways of understanding the existence of social links. Biologists must opt for one or the other and it seems that the explanatory power of Hamilton's theory is greater than that of Wynne-Edwards'. But there is a specific case where both can be related, precisely through the meaning which the functional character of the evaluative term acquires. I consider that in this respect a thesis like the following can be formulated:

The structural sense of an evaluative term like 'good' contributes to the transformation of *kin selection* within the human species into *group selection*.[6]

A family group, in which there exists a reasonable possibility of inclusive fitness giving rise to conduct which would have to be considered selfish from the point of view of the gene, and altruistic from that of individuals related to each other, represents a magnificent situation from which to study the functional meaning of moral phenomena. In the imaginary (and absurd) case that individuals could ask themselves about the advantages of having innate tendencies destined to bring about a conduct which benefited the fitness of other individuals (his relatives) in terms of the overall possibilities of genetic conservation, in that case, I repeat, the answer to questions like 'why should I be moral?' would in no way permit the Kantian distinction between preference and reason. The reasonable and the sentimental of course would closely coincide. So even though this assumption has no sense for reasons so obvious it would be an offence to pause over their explanation, we have no choice but to accept that the role played by moral language at the moment when linguistic capacities are incorporated into the phylogenetic process has to be compatible with the meaning of that profound identity. Only a moral mechanism of evaluative character regarding the situation of symmetry is selectively superior. 'Good' is going to be applied to actions whose consequences are extended to the group through inclusive fitness. There are no breaks in the continuum of expansion of the benefits of conduct; what has to be considered through the critical character of language within the beta-moral level as significant to go on postulating gamma-moral norms in the group (and emic delta-moral ends) will have to be made compatible with everything implied by the presence of evaluative meaning associated with the concept of 'good' under the criterion of symmetry and similarity. That would not be so in the case where the appearance of larger populations, in which non-related groups exist, causes the mechanism of inclusive fitness to lose its value. But when such a thing occurs, man already has traditions in a wide cultural sense which of course include a range of significant moral elements. Already means of identification have been constructed which are genetic (in the sense that they make use of genetically determined aptitudes and capacities) and are not genetic (in that they respond to the appearance of cultural elements determined as much by the ecological niche as randomly). These means of the cultural tradition allow the extension of the character of 'relative' to members of the group who are not related. The system of conversion of families into clans and fraternities appears which has been excessively studied by sociology and anthropology, *Gens* and the gentilique connection appear. Ethical systems such as the Jewish system referred to in chapter 6 appear. It is the moment when the social structure brings to a close what Marx considered an idyllic situation of primitive communism, and this leads us to the complexity of ethical norms developing within tradition. But it is not a matter of proposing that in our groups, or in those of the Asian empires, or medieval fiefs, no fossilized ethics can exist, irrespective of whether in certain isolated cases they might be present. What is essential is to realise that the evaluative structure has already established a deep sense of the concept of 'good' which includes fairness. If symmetry is lost, that which eventually must be considered 'good' has to be justified morally through positive evaluations of an asymmetric class situation, of the implicit acceptance of dependent hierarchicization. Otherwise we have abandoned the field of ethics and placed ourselves in that of random physical oppression, something which, on the other hand, cannot be said to be either rare or exceptional in relations between human beings.

If the thesis about the phylogenetic value of the presence of moral language is correct, if 'good' as fairness is one of the means which could trigger the appearance of social links as sophisticated as those found in human groups, then the very structure of moral language regarding the beta level receives sufficient strength from innate determinants to justify that fundamental *a priori* principle able to show what the asymptotic line of desirable ethical progress is, and in what way in consequence comparisons and preferences between emic traditions with different moral contents and judgements can be made. None of that eliminates the frequent tendency to use hierarchical impositions linked to social class or to any other element of differentiation within groups. Neither does it confer to ethics value to 'ensure' a given future for the species. It only means that if we accept the significant presence of the moral phenomenon in the human species, and if we claim that to make moral judgements is habitual within cultural traditions, we have found an element able to permit us to apply rational means to evaluate codes: an element which essentially depends on the meaning which, for phylogenetic reasons, is assigned to the term 'good' and is conceivable in beta-moral terms. The question of whether throughout the phylogenetic process innate knowledge of some element related to the gamma or delta-moral levels was also incorporated is a very different matter, one which in my judgement gives reason to harbour serious doubts, for the same reasons as lead to the proclamation of the importance of the critical character of language.

In fact the claim of the supremacy of the structural characteristic 'good' as a guarantee of fairness is the basis of one of Hare's requisites for ethical judgements: universalizability. The support which biology can give in this regard is none other than that of assuring through causal explanation that such a line of reasoning is correct, and providing the basis for arguments about beta-moral autonomy. But for such success to have come accompanied by more radical claims of biologization of ethics seems to me to be more the result of mistaken philosophical speculation on the part of certain biologists than the direct consequence of theses which allow one to understand the process of duplication of nucleic acids and genetic evolution.

CHAPTER 9

ADVERSUS LIBERALES: THE RIGHT TO EXCELLENCE AND DISTRIBUTIVE JUSTICE

If modern man finds that his inborn instincts do not always lead him in the right direction, he at least flatters himself that it was his reason which made him recognize that a different kind of conduct will serve his innate values better. The conception that man has, in the service of his innate desires, consciously constructed an order of society is, however, erroneous, because without the cultural evolution which lies between instinct and the capacity of rational design he would not have possessed the reason which now makes him try to do so.

Man did not adopt new rules of conduct because he was intelligent. He became intelligent by submitting to new rules of conduct. The most important insight which so many rationalists still resist and are even inclined to brand as a superstition, namely that man has not only never invented his most beneficial institutions, from language to morals and law, and even today does not yet understand why he should preserve them when they satisfy neither his instincts nor his reason, still needs to be emphasized. The basic tools of civilization – language, morals, law and money – are all the result of spontaneous growth and not of design, and of the last two organized power has got hold and thoroughly corrupted them.

F. A. Hayek, *Law, Legislation and Liberty* (1979)

So far I have referred to the four levels which the moral phenomenon can be broken down into for the purposes of studying its relation with certain causally explicable determinants. In this chapter I am going to abandon this shall we say formal perspective in order to make clear some of the consequences of using a theory of ethics like the one proposed, and their significance for certain theses about the relation between biology and morals as stated by Hayek. The ideological content of determinism in moral matters as proposed by sociobiology has probably had the greatest echo within the critiques and commentaries by specialists from the social sciences, or simply by philosophers. So far I have not referred to these aspects, since as I have already maintained they seem to me to be paradigmatically separable from the body of theory itself and peripheral to its acceptance or denial. Moreover, within the theory discussed here there is a sufficient level of beta-moral autonomy to allow the idea of moral behaviour based exclusively on instinctive impulses to be left aside. By the same token, I must point out that if Social Darwinism today seems worthless, it is due to its scant explanatory value and not because it might lead to fascist-type excesses.

So, it will not be ideological motives which impel me to adopt one position or another in the specific questions I am going to pose, but rather motives coming from the interpretation of evolutionary theory which until now I have used as a background. It is not a case of invoking scientific 'asepsis', but of a need to regard this last chapter as an exercise in the application of the system proposed.

Even so, the ideological bias appears, a little indirectly, because the two topics I am going to refer to (the right to excellence and distributive justice) have been dealt with by one of the most noted representatives of liberal ideology, F. A. Hayek. Liberalism seems to have at least as many supposedly orthodox interpreters as Marxism itself, but perhaps none have known how to take the logical demands of liberal ideology to such clear and radical proposals as Hayek. I will speak then of 'liberalism' and 'liberal ideology', identifying both terms with the area corresponding to Hayek's thinking, albeit at the risk of committing an obvious and crude reduction.

The right to excellence is not a theoretical claim which can be

reduced to circles of academic discussion and its possible significance limited to that sphere. It represents a political alternative
and an inescapable reality, since the budgets which governments
devote to the education of its citizens are usually high, in some
countries at least, and what is being questioned is the use to which
such funds are put. If, thanks to a line of argument based on
biology, we can establish the existence of different intellectual
capabilities in the population and obtain empirical means to
detect at what point on the scale each citizen is placed (by I.Q.
tests or any other method) the investment of similar quantities of
money in the education of each citizen could be a moral error as
well as an economic waste. Liberalism demands differential in-
vestments since different degrees of intelligence exist and so what
happens with egalitarian policies is that some citizens, those of
greater excellence, are denied their right to cultivate and develop
their capacities. I stress that this is not merely an academic
proposal. In the symposium held in Caracas in December 1981
commemorating the centenary of Darwin's death, the question of
the right to excellence was debated in a discussion with the
participation of the Venezuelan Minister for the Development of
Intelligence, who was severely criticized by Hayek himself who
was present. To opt for one policy or another probably represents
one of the most significant alternatives that can be taken today in
the field of state intervention in the well-being or otherwise of its
citizens.

In general, the critique of liberal theses usually reasons through
the ideological defence of egalitarianism, or through a critique of
the empirical means supposedly able to classify people on a scale
of intelligence. I am going to leave both questions aside because
what concerns me is to establish the connection between biological theory and the right to excellence, or, to put matters in their
most radical form, the right to genius. Is there any biological
model able to explain the existence of geniuses in relation to
theses of natural selection?[1] The answer is not at all obvious since
the significance of the theses of biological functionalism is more
complex than it seems. In fact, the previous question could be
better formulated if modified slightly: if intelligence is in itself an
adaptive advantage, why are there not many more geniuses? Why
are we not all geniuses in some way?

It is obvious that this interrogative only makes sense if it is demonstrated that genius, the existence of geniuses, is in fact a selective advantage for the human species. I suppose that to deny it would make most Nietzschians turn in their grave, and that a brief review of the history of humanity could convince even the most hardened and puritannical egalitarianist of the importance of people like Galileo and Newton. I am not interested in arguing for something so obvious, but in examining the relations between intelligence and the biological substratum from a perspective such as Darwinism.

Hayek presents modern society as the result of human activities which have been sifted through competition between different groups. Given that groups did not find identical solutions to problems of survival, and that historically the blatant advantages some of them obtained can be seen, not much effort is required to imagine that some solutions were better, adaptively speaking, than others. Hayek establishes in any case that those practices were adopted for unknown reasons, and perhaps purely by accident (1979, p. 155). What is really important is that institutions, moral norms, and laws which give man selective advantages (in the present sense of higher rates of reproduction over time)[2] had to be acquired, according to Hayek, before critical judgements about such advantages could be made: "*Man did not adopt new rules of conduct because he was intelligent. He became intelligent by submitting to new rules of conduct*" (1979, p. 163). Hayek underlines that paragraph and does well to do so since it is the basis of his argument. But it might be founded on a misunderstanding. The word 'man' in this paragraph is somewhat ambiguous. The behaviour which converted man into an intelligent being (as long as we employ an *ad hoc* definition of intelligence) must be attributed to genera and species related to the modern *homo sapiens sapiens*, but not identical to him. It was moreover probably related less to institutions, positive moral norms, and laws than to other types of activity connected with the search for, preparation and ingestion of food developed by a remote ancestor of ours responsible for the olduvainsian culture, *homo habilis*.[3] If one prefers to define the threshold of 'intelligent behaviour' by a higher hurdle, one would have to mention the proficiency of *homo erectus* in obtaining fire and constructing implements like

achelensian axes. There is widespread consensus among anthropologists that characteristics so closely linked to the human
condition as the aesthetic sense and belief in another world
formed part of the behaviour of a species very different from our
own, the Neanderthal. But even if all this is denied, we still do
not lose very much. The behaviour which caused hominids to
become intelligent appeared four million years ago, or less if we
pay heed to those who propose dating methods through comparative analysis of the molecular structure of protein, but in any case
it is within that scale. New rules of conduct turned the *hominid*, if
not into man, into an intelligent being. And our species *homo
sapiens sapiens* already had the genetic endowment necessary to
make discriminatory responses to a complicated and changing
environment using intelligent criteria. Man was able to go on
increasing the complexity of his surroundings by incorporating
new significant elements through a different evolution, cultural
evolution. And he changed his rules of conduct. He adopted new
norms and got rid of old ones. At times those changes were
gradual and successive, at others sudden, and difficult to explain
from our perspective. But in no way could they serve as significant elements for genetic evolution if what we are referring to is,
as I suppose it to be, what happened from the neolithic revolution
to the present day. Quite simply, not enough time has passed.
Thus the rules, laws, and institutions which have cascaded after
one another since man constructed towns, became a peasant and
domesticated animals can only have made use of a genetic inheritance which was already established in the long dawn during
which ancestral hominids collected roots, stalked, hunted and
just managed to survive thanks to the progressive use of intelligence and language.

But these specifications do not seem to add much to the
clarification of our problem. Let us re-frame our initial question.
How, since neolithic times, or even before, do human groups
break with previous laws and customs? It is a thorny and awkward question. Man secures his survival within the group; his
personal intellectual faculties do not carry sufficient weight to
make individual adaptation to the environment possible. That
does not mean that hominids established anything like an original
contract convinced that life in a group was good for them.

Perhaps some of them, if they ever reflected on the matter, opted for the opposite idea and separated themselves from a group which demanded a lot in return for membership. We do not know if, intellectually speaking, they would receive sufficient personal gratification through the life of a hermit, but we do know very well that when they died their adaptive formula died with them. Only groups united by norms of conduct which demanded a genetic tendency to favour submission to and acceptance of authority, at least regarding the years of infancy and early adolescence, survived. All success in the human adaptive task had to be based on the value of traditional institutions. Hayek insists on this: all progress must be based on tradition. And, what is progress? From the historical point of view it has been a change in the behaviour, values and the very institutions of a group. Let us leave aside whether we have to accept the positive or negative character of a change in that sense. It has always been achieved in this way: within a group strongly bound to traditions, one or several innovators emerge. Someone with a brilliant intuition – or if preferred a lot of luck, or divine inspiration, or a delirious illness – conceives a new type of behaviour which in the long run changes everything. While the group is building the new and firm tradition, change will certainly not be noticeable. But without the heretic, progress would not have come into existence. There is no doubt about the meaning of Hayek's formulation: while *everyone* remains bound to the rules of tradition there will be no change. Individual transgression is necessary because it is unthinkable that *everyone* simultaneously – or even a large number of the members of a group – will suddenly be converted into apostles of the new idea. Little matter if the heretic pays for his daring with his life if the seed of the new order is scattered throughout the group; and little matter if the process of transformation is different from the one Hayek presents in the sense that we might give greater importance to collective or peripheral elements or propose the need for objective conditions vital for change. This will only mean that of the many innovators who emerge, only a few will make their ideas valid in the long term, a different question from the one which now occupies us. What is significant for our purposes is that *all* changes have, perhaps in their remote origin, a heretical innovator.

Or a genius.

What is a genius?

For our purposes, to define what a genius might be is a little embarassing. We are too used to a sense of genius which, moreover, we extend to fields which are difficult to compare to each other. A musical genius obviously does different things from a mathematical genius, or a genius at chess. All of them however obtain this qualification through comparative criteria. It does not seem to bother us greatly whether the level of mathematical knowledge of a graduate of any of our universities might be very superior to that of Leibniz or even Poincaré himself. The discussion of the restricted principle of relativity at present occupies a great number of people, but none of them receives the qualification of genius for this particular activity. A genius is without doubt relative to his time and his fellow citizens. Furthermore, the everyday use of the term has allowed its meaning to be extended into fields where the question of genius would initially have been very suspect. Perhaps it is in a metaphorical sense that we can speak of geniuses in arts such as killing, by dint of taking questions as thorny as that of the rationality of the means to the limits, and recognising in an assassin the unsurpassable use of elements within his reach to kill people. But there does not seem to be any metaphor in the expression which classifies as a genius the chess master who in everything else is an individual of such low mental level that he has difficulties in undertaking even the most usual domestic tasks. Probably the latter would be defined as a genius only in a society which allows the survival of a person unable to distinguish between what is edible and what is not. If we leave these extrapolations aside and accept the general thesis that genius is a relative characteristic, we will have to accept that one is a genius depending on a standard, on an average which, if we limit ourselves to the field of intelligence, is marked by the intellectual level of the group in which the genius appears. But of course the matter is not so simple. The intellectual content of human groups measured according to the standard of modern Europe has undoubtedly been growing. The average Spanish citizen of today knows many more things than the citizen of the Spain which received news of the existence of a new American continent. Does that mean that a modern genius requires greater capabilities?

Let us avoid confusion between two completely different

things: the capacity for genius and the ultimate result of the intellectual level of a human being. The final level is the product of the interaction of two groups of factors, of which only one corresponds to the genetic inheritance of the individual. His innate capacities will be completed in the years of infancy through the increase in the complexity of his brain in the process of language acquisition. And, moreover, the group in which he lives will mark out some very precise limits and conditions through which the direction of the intellectual process will be defined. We will never know how many shepherds could have been Einstein if they had had the necessary means to study mathematics. I believe very few, but I will not argue with those who insist on the hypothesis that the number would be very great. What I am certain of, and what any reasonable person will have to admit, is that Einstein's genius was something exceptional. It is obvious that not everyone who has had the social resources Einstein had at his disposal, nor the greater number of people who historically have found themselves in conditions similar to the geniuses of their time, has in turn become a genius for this reason. Let us accept then that it is essential to have something peculiar to genetic matter itself, something innate which makes the emergence of genius possible if the social environment permits. Something which not everyone has, and which for my part I believe to be very rare. So today, in an intellectually advanced world, do more geniuses exist than in the intellectually more backward world of the neolithic period? It is an extremely difficult question to answer because of the need to take account of a reference criterion. Let us suppose that we can in any case decide that it is possible. The reasons we adduce in favour of our verdict could be that there are greater opportunities for developing genius, flimsier institutional barriers, the loss of superstitious fear of knowledge, etc. All that will only affirm that the second part of the condition of genius, social and cultural surroundings, is different. And what can be said of innate capacities? Today does the human species have a greater number of potential geniuses per generation in statistical terms? The answer must once again be negative, since we are postulating the impossibility of having significant genetic changes in so short a time. In a neolithic period there would have been a similar number of beings capable of seeing the world with a different critical spirit, if that is what we reduce genius to, in spite of none of them having studied at Harvard, received the Nobel Prize or composed fugues for the organ or clavichord.

It was necessary to accept that in order to discuss the basic problem posed by the existence of geniuses. If the existence of genes providing the capacity to view the world and understand it in a different way is something which is adaptively valuable, how is it possible that this type of gene has not spread throughout the whole human species? Why are there not many more potential geniuses? Why are we not all potential geniuses? There are only two possible answers from evolutionary theory:

(a) The question is mistaken. We are all in fact potential geniuses.
(b) The evolutionary stable strategy of the human population rests on having few potential geniuses in each generation.

Those who choose the first response will have to use strong reasons to deny the existing evidence against a proliferation of potential geniuses. In a world where opportunities for education are growing, at least in some societies, the only thing that appears to be growing in proportion is the difference between the elite of thinkers and scientists, which, however much we relax the concept of genius, is small, and the wide world of people anchored even proudly and defiantly in their normality.

But to simply accept the second option is extremely risky. Why is it better to have only a few potential geniuses than to have many? And in that case, how does the process of selection manage to keep the rate low? Perhaps the response to the first objection gives a lead to finding the very risky solution to the second. A typical answer to why it is better to have few geniuses than many is usually given by taking advantage of the tautologous character of the definition of *good* in Darwinian naturalism. All existing characteristics are the best because they have been selected for that reason, thus the answer to why it is better to have few geniuses than many lies in the fact that it has happened that way and so it is without doubt the best way. But I imagine that the question will be granted a higher heuristic value and that circular arguments will not lead us very far. It is necessary to opt for another type of answer which is moreover detectable in the way the selection of progressively intelligent conduct could be produced in species of competing hominids.

We will have to accept right away that the appearance of especially critical behaviour among hominids subjected to a process of change in their eating habits had to be an extremely useful element. The recognition and association of such things as climatic conditions, the habits of animals, both predators and herbivores, raw materials for the production of scrapers and knives, and the organizing of hunting and gathering strategies are all matters which would certainly lead to the success of those groups which had a progressive number of technical innovations. But technical innovation, like any other behaviour change is, let us remember, a final result. The appearance of individuals potentially gifted with genius would not be at all significant for the process, and in consequence, for nothing which natural selection could impose in the long run, if it were not a case of a phenomenon suitable to the environment in which members of the group find themselves. And that environment contains not only animals, plants, rains, droughts and rocks, it is formed primordially of individuals of the same species as the potential genius, in whose company he has to grow, survive and reproduce. This ecological niche initially represented extremely precarious circumstances for the survival of the group. But if that group has managed to maintain itself until the very moment at which we are imagining the appearance of the genius, it is thanks to the power of tradition. The significant element in the process of adaptation through critical conduct is, over extremely prolonged periods, tradition. Exceptionally the presence of a genius introduces possibilities of change, probably joined to other factors external to the group (such as climatic changes, migrations, or general modifications in environmental conditions) but which remain a mere possibility. Much more exceptionally a change in tradition will come to fruition through the appearance of a genius who is especially able or fortunate in the set of conditions. The group in question will have better ways of transforming its environment and, this being possible, those techniques will extend beyond the physical limits of the population to lay down roots in other neighbouring groups. An exceptional and strange break in tradition will have come about.

But it is tradition itself which is responsible for the triumphant strategy of the group in almost all imaginable circumstances, leaving aside those minimal exceptions, however significant they might be in a perspective like our own where we can pass over hundreds of thousands of years without noticing anything other than exceptional events. If the balance between tradition and innovation were modified slightly, that is, if the percentage of potential geniuses were slightly altered, the group would disappear. Survival would not be possible for a group of hominids who systematically undermined traditional behaviour and norms, because it is strict and severe subjection to that conduct and that norm which provides the basic securities which from time to time will allow change. The world of hominids could barely tolerate the proliferation of heretic geniuses, just as our own world would find it difficult. The value of the exceptional is as closely linked to the condition of being exceptional as our aesthetic sense can indicate. Le Corbusier used to say that when he strolled through a city he would find some silent buildings and others which suddenly seemed to be a symphony of music and colour. But I doubt whether Le Corbusier, with his brilliant sensitivity, would have tolerated a city constructed entirely of outstanding buildings which emitted what would be a terrible cacophony of sounds and senseless colours. One house by Gaudí in a city like Barcelona is a poem; a whole city designed by thousands of architects like Gaudí could become a confused nightmare.

But if we accept that the evolutionary stable strategy of a community of hominids would be that of a population with a very meagre content of potential geniuses and a very high content of individuals who greatly respect the traditional way of interpreting the world, we still have to explain how it is possible that the proportion would be exactly that maintained in the gene pool of the species in question. The alternative to this interpretive requirement would be to consider the previous response correct and resort immediately to the mechanism of group selection: those groups possessing the proportion we consider to be ideal would be in a better position to survive, thanks to an adequate degree of respect for tradition and the presence of some individual geniuses who, when other environmental circumstances permit, would be the catalysts for change. However, genetic biologists usually scowl when group selection is mentioned. Ideas like the 'gene pool' remain an abstraction and, however much groups might have different chances of survival, strictly individual phenomena rule when it comes to transmitting characteristics to the next generation. I suppose I will not be allowed to advance the idea that the group consciously devotes itself to the selective breeding of potential geniuses.

However, that is what they do, in a way which is perhaps only slightly conscious. Given that the individual conduct of the potential genius has to be translated more or less strongly into a critical attitude towards the traditions representing the standard way of life in the group, the weight of tradition will tend on its own to project a more or less pronounced form of heretical stigma onto the individual who is different. If the stable evolutionary strategy of a group in competition with others assumes sporadic genius as an advantage the evolutionary stable strategy of the individual in competition with other individuals within the group dominated by tradition does not appear to reward over-critical behaviour too highly. Rather the reverse; the life of a genius, even nowadays, would not merit the admiration of a utilitarian set on drawing up a balance sheet between pleasure and pain. Our own groups are still used to punishing severely those who are different, except in extremely rare cases. Perhaps the opportune question should be 'how can geniuses exist amid such difficulties?' And the answer cannot be reduced to stating that the selective advantage that they do exist is so great in the long run that it is not odd that our species has individuals able to break with tradition now and then, however much the genes of potential genius are scarce and their individual carriers habitually unfortunate. That would mean falling once again into an argument about the mechanism of adaptation which would be based on group selection.

To be able to support a theory of the type proposed here, it will be necessary to explain, by contrast, what type of selection can favour individual transmission of characteristics in proportions such as those which we take to be good: an abundance of average intellectual levels and scarcity of potential geniuses. How, if a potential genius in fact has so many personal problems, do genes of this type manage to survive in the gene pool of the species.

Perhaps through neutral selection? How then does potential genius turn out to be so evenly divided geographically?

To answer these doubts it is necessary to understand another aspect of potential genius. Until now we have moved along a line of argument in which what was valued was the capacity of the potential genius to provoke a movement under certain environmental conditions (or if preferred, historical, or both; it all depends on what we understand by environmental) which will lead in the long run to the whole group changing its traditional conduct, and I have mentioned in passing that this potential must include things like a different way of interpreting the world, at least in certain partial aspects. Perhaps that image is too restricted by what today we consider exceptional intelligence. The submerged potential genius in, shall we say, the cultural surroundings immediately prior to the olduvaiensian traditions not only had to have sufficient perspicacity to anticipate what would happen if concave forms of flint were taken and repeatedly beaten until a lengthened cutting edge was produced. He also had to attain sufficient synchronization between mind and hands to be able to do so. In the period referred to, manual dexterity had to be much more decisive than among ourselves, for reasons easy to understand. Perhaps other connections with senses like taste and smell were also of extreme importance in obtaining specific data on strategically superior conduct. And all those elements not only meant a decisive advantage in their time for the group: we are practically enumerating the virtues which would convert a hominid into an individual apt at ensuring his own survival. What today could be considered as scientific experimentation and renovation of paradigmatic elements is situated in areas relatively distant from the everyday task of survival (if we take for granted the perhaps overly-optimistic thesis that a scientist can survive without submitting himself to utilitarian criteria regarding the value of his work). But in the Upper Palaeolithic such experimentation occurred precisely along the same channels as the search for food, food which was perhaps sufficiently different to suddenly allow new, and hence abundant sources to be counted on.

According to these premises, genius behaviour would move between two contradictory norms with opposite results. Novelty would imply some continual advantages, probably significant however small they might be, and of course proportionally high in that their discovery represented new sources of food. But each novelty would also mean the transgression of cultural norms fixed by tradition, and consequently a danger for the group institution as well as for the very security of the potential genius. How points of balance between both extremes of theoretically possible behaviour (absolute innovation with transgression of all norms and rigid subjection to traditional values with a curb on any progress) could be found moreover depend on data from the ecological niche. In any case a genius potential is possible which would include elements advantageous for personal adaptation and in consequence for genetic conservation of intellectual and manual superiority as long as the advantage were not so evident that it provoked excessive institutional disquiet. I think the idea of long periods in which genius would only be advantageous on the condition that it was numerically reduced can fit quite well with data we have on the very slow process of transformation of oldowan implements into achelesian ones, and the no less prolonged lapse of predominance of cultural tools associated with the appearance of *homo sapiens sapiens*.

This way of seeing things implies moreover a very different moral from that which Hayek draws from his own position. For Hayek the (modern) tendency towards egalitarianism as a political and cultural strategy is an immense error. Given that human progress lies in that glorious difference of genius, the right to excellence is the supreme value of the human species. If there is any doubt about whether this is really his position, anyone who attended the symposium commemorating the centenary of Darwin's death, referred to earlier, could dissolve it. There, an explicit discussion about the assignation of state budgets to different types of education programmes was held, and the line of thinking headed by Hayek clearly stressed the error in trying to give similar education to the whole population. This world, for him and those who think like him, is a world of geniuses, and if one must opt (as is usual) either for a similar level of average education for a large number of citizens or elite education for the few, the theory of natural selection shows that the second is the only valid solution. The other would even be suicidal.

I hope that at this stage the reader is able to distinguish

between what could be called the general level of education which up till now has been growing in broad terms in European history, and genius differentiation. An education programme designed to promote genius, from this point of view, is not necessarily morally mistaken – something which the theory of natural selection could not go into except as an accessory element under risk of tautology – but would, in any case, be economically useless in Hayek's proposals. The process which he describes is, let us recall, random and beyond any possibility of rationalization. We cannot then predict and shape the type of education for future geniuses, because we do not know what lines of rebellion against current values, norms, and behaviour will emerge. That could only be done from a rational perspective for interpreting progress and I fear that such a perspective would quickly meet with obstacles in carrying out programmes of that type; at least to me that is the thesis which Hayek seems to support. And if we were just to accept the random process and defend an elitist education made by chance? Another waste. On what criteria would future geniuses be selected? Those with a greater intelligence quotient? Unfortunately we are only able to measure such things by our current standards, the very standards which the future genius would reject. And if it transpires that we are forming elites who are progressively respectful of our traditions? The best way to ensure a breeding ground for future geniuses seems exactly the opposite: give the whole population a sufficient level so that the potential genius can show him or herself. The rest will certainly come on its own.

The second of Hayek's proposals of liberalism that I should like to discuss refers to the use of state apparatus to impose, through corrective criteria, a principle of distributive justice. The refusal to accept a redistribution of incomes for example obeys a thesis which is at least partially based on biological considerations to classify as mistaken all types of intervention in processes which, like the economic process, are 'natural' and are better resolved left to their own dynamics. For Hayek, the liberal policy which is able to ensure that society functions in the best way is the most radically understood interpretation.

In reality this thesis takes two forms which could be considered historically successive if it were not for its recent reappearance in liberal doctrines like Hayek's. Initially, and following Locke as much as Hutchinson and Lord Kames, the question most closely linked to growing liberal ideology, that of private property, belongs to the field of natural rights. The fruits of personal labour are of concern to human nature and in consequence any limitation on free private possession works against the laws of nature. In its later form, the justification of liberalism takes on a functionalist note: it is, in sum, the only means of achieving things like social peace, wealth and well-being, affairs in which the fact that such a system compares favourably with State economic intervention is pointed out. If these two are combined into one, it is through F. A. Hayek's particular exposition of functional theses, drawing support from sociology and the innatist theory of evolution to introduce once again natural reasons as a basis for that functional character.

The functionalist thesis is tempting, above all in its comparative analysis of State systems which lead to oppressive regimes with scorn not only for individual liberties, but for what could be understood as human dignity (assuming that the counter-argument about what type of human dignity is desirable is not going to be accepted). It usually commits the methodological clumsiness implied by the rejection of a given interventionist theory by making use of empirical examples, but aside from those details, it will be useful to focus on the proposal in hand, which is the critique of the liberal naturalist thesis, onto which we have still not projected any suspicion of fallacy.

We could immediately point to a paradox. How can a universal characteristic (genetic in the final instance) contain strategies especially directed to the limitation of its universal application? The concept of private property cannot get rid of its negative sense, that is, of the implication that it deprives others of use, without risk of failing. As Parsons had to subtly point out, that implies that private property and power are terms intimately connected. There is only private ownership of what is scarce and desirable, and being scarce and desirable it is necessary for others. By claiming what is mine I demonstrate my opposition to allowing any other member of the community to use my property. And such well known sayings come, I stress, from the structural-functionalist approach itself; in no way are they the result of applying axioms from paradigms opposed to the liberal tradition.

Under these circumstances we will have to explain how something which is claimed to be of a biologically universal nature, and refers moreover to things of scarce and limited character, can rest indefinitely on particularization. Let us suppose that the analysis must be applied to goods susceptible to privatization such as cultivable land in an agrarian social group. If to obtain the right to private property it is necessary to mark out and remove part of the total amount of available goods, it is obvious that we can without any difficulty propose a situation in which the exercise of the right to property necessarily supposes the privation of other group members. The so-called universality of law is thus at least restricted by the division of the group into two types of citizen: the have's and have-not's: and that quite simply means transgressing the principle of universality. But universal principles do not bear exceptions well. If we already have a group necessarily deprived of its natural rights what reason can there be for frustration being prevented from spreading from everyone?

It would be stupid to claim that liberalism merely ignores the difficulty. In fact every liberal theory is necessarily linked, as is known, to the development of contractualism, and the hypothesis of the social contract tries precisely to resolve that conflict from the start. But contractualist theses imply a very tangible modification to the initial idea. Even when there is no immediate logical need (apart from that already pointed out) contractualism (at least modern contractualism) establishes itself along a line of thought far removed from ideas of universality and natural law to attend instead to matters such as the foundation of justice and the development of positive law. In such conditions the response to the previous question takes on a utilitarian note: it is preferable to limit access to property to a reduced group for reasons of collective well-being and functional social organization. Let us leave Hayek's evolutionary alternative until later.

The consequences of abandoning the thesis of natural principles are somewhat dangerous for the liberal position. A slight discrepancy in the desirable objectives of well-being and human dignity would be enough to proclaim the superior functioning of state-run regimes. In general the problem is usually avoided by pronouncing the irrevocable goal of formal individual liberty, identifying it for practical purposes with indicators like the absence of censorship and democracy in a strict sense – freedom of political parties, free elections, self-governing parliaments etc. Unfortunately the strict application of neo-liberal premises for radical and non-interventionist privatization seems to be suspiciously linked to political systems very distant from such an idyllic form of freedom. Remember quite simply the application of Milton Friedman's programmes to countries like Chile and Argentina; and the United States under the Reagan programme. They might satisfy a liberal who is not too demanding in matters of formal freedoms, but it would be somewhat picturesque to define such State systems as absolutely faithful to the principle of non-intervention.

However, let us avoid falling into the temptation of empirical refutations. Where is the fallacy in proposing the advantage, for utilitarian reasons, of an economic regime of free private property in a strict sense? Well, quite simply, to my way of thinking, it would be contradictory to the parallel demand for the fight against an interventionist state in other areas of social life. But this is a question which is too far removed from the topic we are trying to discuss.[4] Let us continue with the previously abandoned question of Hayek's interpretation of Darwinian evolutionism in its new sociobiological synthesis.

In the third volume of *Law, Legislation and Liberty (The Political Order of a Free People)*, Hayek devoted the last chapter to a search for the sources of human values through a critique of Wilson's sociobiology and Pugh's theory of the biological origin of values. I believe the critique as such does not manage to dent the solidity of sociobiological theses, and that in fact Hayek's proposals are quite compatible with the idea underlying Wilson's whole text (which on occasions is explicitly expressed) of culture as a phenomenon immersed in the interrelation between genetic information/ecological niche, with specific virtues for the survival of those groups best placed to find effective solutions to the problems of adaptation. Both Wilson and Hayek claim to follow a Darwinian line of argument and consider that the capacity to learn is in itself innate. For his part, Hayek's rejection of the idea that culture is the product of rational designs does not seem to worry the father of sociobiology too much. But what is significant for our purposes is not what is correct or false in critiques of Wilson, but the way in which Hayek develops his theses. For him, there are three sources of rules capable of guiding human behaviour:

- innate rules established in the genetic code;
- rules, which shall we say are residual, the product of traditions acquired by going through successive types of social structures;
- rules deliberately adopted or modified to serve known proposals. (pp. 159–160)

The scale is hierarchical. Innate rules would be solid and difficult to modify, while the rational rules of the last level form a thin level in comparison. Together the set of rules is contradictory and conflictive and on this stormy characteristic rests the tendency towards accelerated social change. No one who appreciates the ethical formulations of Eibl-Eibesfeld's ethological theory, for example, would be in the least surprised by the panorama Hayek offers.

But the conclusions he draws are indeed original. They begin once again with a commonplace from the ethological and sociobiological perspective: conflicts introduced by evolutionary change which represents passing from living in small groups of individual acquaintance to the Popperian 'open society'. In general, Eibl-Eibesfeld tends without very profound explanations to postulate that man has mechanisms of an innate character compatible with both types of social group, that of direct acquaintance and the abstract multitude, thanks to the use of rituals which ensure personal identification with a large group. It is Hayek who stresses the significance of change.[5] Small societies have innate emotions whose gratification works in favour of group survival; the intersection of their innate rules and the social characteristics of the group makes emotions and ends compatible. But those individuals who are not subjected to compatibility find solutions to make the group evolve from the social point of view. At the end of the process we find an open society in which the rules are abstract and the signs impersonal, and where clear identification with specific ends is not possible. Things like language, morals, the law, and money are the result of spontaneous growth through the transgression of codes, and in no case the result of rational design. We might consider this theory still compatible with the thesis of the social contract, because it is abundantly clear that the explanatory value of the latter lies in its being the foundation of rules of justice and not in any claim to being a system of genetic explanation of the social mechanism. But it turns out to be totally incompatible, as Hayek is careful to indicate clearly, with programmes of so-called 'social justice'. For good reason Hayek devotes the entire second volume of his trilogy to denouncing 'the mirage of social justice'. I find the formulation given here by Hayek exemplary, where he states that the fundamental question arises of whether there exists a moral duty to submit oneself to an authority which tries to coordinate the efforts of the members of society with a view to creating a particular model of distribution which could be considered just. (p. 115)

I imagine that at this stage it is easy to anticipate that Hayek's answer is negative. Even his critique of Stuart Mill rests on the conviction that to include among the meanings of justice, as Mill does, a situation which can depend on deliberate human actions, opens up a way for social justice leading to authentic socialism.

The key to Hayek's rejection resides in the inapplicability of the concept of justice to the results of spontaneous processes. The concept of social justice has no meaning in a market-based economic system (which is what has caused social systems to evolve, precisely by breaking down the marriage of emotions and ends) and what is more, the market order cannot subsist when in the name of social justice, or any other goal, any mode of fiscal policy based on an evaluation of the merits and needs of different individuals or groups is imposed by an authority which has to power to enforce its own criterion.[6]

Such irrationalism might seem monstrous (or adorable) but in any case is very coherent with the principles of liberalism, as long as they are followed with a certain rigour. Let us remember three very significant points:

- Social evolution rests on rationally unpredictable, chance initiatives and constructs rules non-rationally.
- This line of progress eliminates the idyllic primitive situation of identity between emotions and ends.

- The search for rational criteria of distributive justice leads to the breakdown of liberalism (as linked to market economy) and ultimately to socialism.

There now emerges a question to be added to the certainly interesting ideas of Hayek. Why need it be desirable to continue along this line of social evolution? What if it turns out to be better to abandon it and substitute another for it, linked to planning criteria of distributive justice, at the cost of the strict economic formula of the free market disappearing? I suppose that it is obvious that this question cannot be answered from Hayek's viewpoint, unless we resort to the sense of evolution in the context of Darwinian theory. That will be *desirable* if it turns out to be an evolutionary stable strategy of adaptation to the environment. The tragedy for liberal theses is that a very clear example of a triumphant strategy in their sense exists: the totalitarian State.

What if such a social system reduces individual liberty and human dignity to nothing? Well, who cares? We are following a line of argument in which individual emotions (which should include all these matters, as did the Scottish Moralist school) have ceased to be gratified by their identification with ends and values. There is no such thing as 'natural goodness' in a civilized society. If Hayek rejects economic rationalization it is not because he is critical of such losses (which in any case he considers necessary for evolution) but because it seems to him absurd and childish to design an economic system rationally *"We are not intelligent enough for that"*[7] But what if some men inclined to contravene the established order of the open society insist on doing so? The result will be economic chaos, as their intelligence will not be up to such an enormous task. It is truly distressing that we cannot know what will become of a system like that of the Soviet Union in the long run, but I should like to venture the reflection that, as Hayek says, *any attempt* to arbitrate a programme of distributive justice, even the mere objective of assigning specific portions of the social product to different individuals or groups (Vol. II, p. 115) undoes the liberal model and leads to socialism. When it comes to utopias, the liberal vision is unparalleled.

NOTES

CHAPTER 1

[1] Even though, without too much distortion, we can concede that contemporary moral philosophy is in a significant sense the analytical philosophy of moral language, it is also true to say that for some time there have been symptoms of a tendency towards a breath of fresh (or at times not so fresh) air. The first sign in this direction of which I am aware (apart of course from the now old postures of American pragmatism) is in an article by Miss Anscombe 'Modern Moral Philosophy' (1958), where she laments the lack of answers which both classic and 'present day' ethics provide to questions about what constitutes a good man or a just action. The clearest example of this new attitude is the theory of virtue anticipated by G. Warnock *The object of Morality* (1971) and developed for example by P. T. Geach *The Virtues* (1977), James Wallace, *Virtues and Vices* (1978), and Philippa Foot, *Virtues and Vices* (1978). Sociobiology implies a body of theories connected of course to a work such as that of Wallace and could be seen in this sense as an alternative capable of satisfying Miss Anscombe. Even so, I still believe the analytic paradigm can be taken to be extremely widespread, if not predominant.

[2] In Chapter 2 there is a commentary on some aspects of Lumsden and Wilson's work which are significant for the purpose of this book.

[3] It is not absolutely fair to suppose that authors such as Midgley or Singer accept the reduction of ethics to simple sketches. What I mean is that their critiques follow sociobiological propositions closely, whereas my intention here is to change the whole approach to moral phenomena. In fact biologists who are much more sensitive to the philosophical side of their discipline, such as Francisco J. Ayala, have realised how complicated moral action is, so it is perhaps superfluous to emphasise the complexities since it has now become urgent to propose ways of solving them.

[4] See S. Toulmin, 'Human Adaptation' (1981) on the various concepts of 'adaptation' and its relevance to the explanation of human questions. I assume my own position will become clear throughout this book.

[5] The concept of inclusive fitness is probably one of the central axes in sociobiology and is thus omnipresent in all the specialized literature. The reader interested in calculations about the genetic structure of populations in order to measure inclusive fitness (beyond the elementary suggestions made in Chapter 8) can refer for example to Peter O'Donald, 'The concept of fitness in population genetics and sociobiology' (1982), an article which also includes a formulation of the kin selection model; or to Richard E. Michod, 'Positive Heuristics in Evolutionary Biology' (1981). J. A. Stamps and R. A. Metcalf, in 'Parent-Offspring Conflict' (1980) analyse empirical evidence on parent/child relationships in terms of genetic fitness, and put forward the conflict model as an

alternative to the altruistic model. B. J. Williams in 'Kin Selection, Fitness and Cultural Evolution' (1980) criticizes functionalist interpretations of cultural evolution and the very concept of inclusive fitness in less fertile species. The question of inclusive fitness and the two models of group versus kin selection will be returned to in Chapter 8.

[6] This has been denied by authors such as Bertram (1982) and is, in general, difficult to defend from the perspective of 'inclusive fitness'. Trivers' thesis about reciprocal altruism favours the group selection model. See Chapter 8.

[7] The Prisoners' Dilemma can also enlighten us about an interesting aspect of altruism. Peter Singer advises his readers that the really favourable strategy is that of the altruist, a consequence of applying decision theory *ad hoc*. But, aside from the type of strategy which might be *best*, which would be the *real* one? To what extent would two ordinary human beings subjected to this dilemma opt for the altruistic solution? The answer is very difficult, because of the need to call on statistical terms as much as for the variations which might be introduced by the punishment/reward relationship, and also the magnitude of the greatest possible loss. An experiment which is perhaps not too relevant in scientific terms, but which is at least illustrative, was carried out by Douglas R. Hofstadter, who offered alternatives similar to those of the Prisoners' Dilemma to twenty of his friends (among whom figured specialists in games theory, such as Martin Gardner and Robert Axelrod) and made them choose between altruism and egoism. The results, which I repeat should not be taken more seriously than the author intended, showed 14 egoists to 6 altruists, with alternatives which meant winning U.S. \$57 if they all chose the altruistic solution, \$19 if they all proved selfish and a variable amount depending on the number of altruists and egoists in other cases (each player who chose egoism received \$5 for each altruist and \$1 for each selfish mate; if he preferred altruism, received \$3 for each altruist and nothing for each egoist). Personally, I believe Rawls is right to consider that maximization, if applicable to human behaviour, both economic or otherwise, (which Rawls believes it is and I do not) is governed by the minimization of risks, above all when one's own survival is at stake. Perhaps this supposition could increase the percentage of altruists in a real situation and in Hofstadter's game, where nothing of great value is to be won or lost. Hofstadter's article contains a bibliography about the Prisoners' Dilemma (1983).

[8] Two similar traits in two organisms belonging to different species can be so through homologous similarity (the organisms are genetically close and descend from a common ancestor who probably developed the trait), or analogous similarity (as a result of different but convergent evolutionary changes). Stephen Jay Gould quotes as classic examples of homology the front extremities of people, dolphins, bats and horses. They look different and serve different functions but they are formed by the same bones. By contrast the wings of birds, bats, and butterflies are frequently referred to as a standard example of analogy, but no common ancestor of any pair of these had wings. (1980b, pp. 205–206)

[9] This example will be commented on more fully in the next chapter.

[10] Francisco J. Ayala considers that none of the three conditions necessary for ethical conduct (ability to foresee the consequences of one's actions; ability to

assess actions as right or wrong; ability to choose between alternative types of action) exists in animals, which therefore do not exhibit moral behaviour. Ethics would appear once an 'evolutionary threshold' had been crossed, which the species approaches only gradually, but passes suddenly (1980, pp. 172–175). The threshold phenomenon, to which reference will be made later when dealing with Chomskyan theses on language, is also known as the 'emergence process' and represents an essential element in any hypothesis of discontinuity between species close on the evolutionary scale. See note 11 to Chapter 4.

[11] In *The Biological Origins of Human Values* (1978), George E. Pugh distinguishes between 'primary' and 'secondary' values present in any system of decision-making which is similar to the human brain. 'Primary' values define those ultimate criteria for decision-making introduced into the system by the designer (in the human being they are, according to Pugh, innate values established in the species during the process of evolution). 'Secondary' values are developed by the decision-making system itself as a practical aid to the problem being solved (in the human being they include moral principles and appear both individually and culturally as an extension of the most basic innate human values (1978, pp. 6–35). Only an evolutionary mechanism can cause primary values to change; these are therefore immune to any rational attempt at 'improvement'. Pugh's is one of the most sophisticated attempts at providing a model for genetic determinism of human conduct until the joint work of Lumsden and Wilson appears. His distinction between the two types of values, however, has no direct bearing on the levels proposed here, except in the sense that his ultimate criteria must be considered as one more delta-moral proposal, with absolute pretensions, providing of course that we accept as 'absolute' a characteristic which has been determined through evolution and which may undergo change in the same way.

For his part, José Luis Aranguren (1958, p. 49) has emphasized the need to distinguish between 'moral as structure' and 'moral as content', following Zubiri. Neither is there any correlation between this proposal and mine, except the common concern to point out differences within the complexity of the moral act.

Ernesto Garzón Valdes has made me take note of the essentially descriptive character of the gamma-level, and as such, of the need to include within it *part* of the task of justification (specifically that which could be considered *emic* as opposed to *etic* according to the criteria explained in Chapter 7). This is useful specification which among other things makes the interrelated character of the levels clear. But the complete (up to a point of course) normative sense of empirical norms as opposed to the open-argument aspects which characterize beta-moral contents goes against this idea. The difference between *emic* and *etic* is, in this context, of secondary relevance. The beta-moral level itself has aspects which are both structural (the sense of the concept 'right' as impartial) and empirical (the very fact of the dialogue, the importance of which no reader of Habermas or Apel could deny). In the idea of maintaining the paradigmatic separation of levels it will now be the empirical aspects of this level which lose relevance compared to the structural ones.

CHAPTER 2

[1] Throughout these pages I shall confine myself almost exclusively to the theme of Darwinian ethics contained in *The Descent of Man*, principally in Chapter IV of the second edition, and within this to the last four epigraphs. Since the article by Robert Richards, 'Darwin and the Biologizing of Moral Behavior' (1982) it makes no sense to insist upon what he there refers to as the early formulations of moral sense in Darwin's work and its relation to Martineau, Mackintosh, and Abercrombie. However, compare E. Manier (1978) chapter 6 in 'Skepticism, Romanticism and the Moral Sense'. Except where indicated to the contrary, all quotations from *The Descent of Man* are from the Modern Library of New York edition, after the second edition of 1874.

[2] This chapter appears in the first edition of *The Descent of Man* under the title of 'Moral Sense' and before the one entitled 'Comparison of the mental powers of man and lower animals'. From the second edition on (1874) it comes after this chapter and as a second part of it, but in the page heading it retains its old title of 'Moral Sense'. The main differences in the text of both editions refer to the critique of utilitarianism which will be discussed later.

[3] All these authors are summarized and criticized by Alexander Bain *Mental and Moral Science. A Compendium of Psychology and Ethics* (1868), a work which Darwin quotes several times in his text. The 'motive for action' in its current form is based on the biological theory of altruism. See Peter Singer (1981), p. 43.

[4] *Op. cit.* 1872 edition, pp. 280–281. The mechanism of sympathy which Bain describes, after Adam Smith, speaks of the indirect hedonistic satisfaction of he who experiences sympathetic sentiments (chapter XI, *Sympathy*, epigraph 9). Darwin has no need of this type of causal explanation of motives for action; as he considers sympathy to be an instinct, its force is sufficiently guaranteed. See *The Descent of Man*, note 21, p. 478.

[5] See Robert J. Richards (1982) p. 57. In any case Darwin recognises that motive and criterion may be combined (*The Descent of Man*, note 42, p. 490), which is not at all strange considering the way he ends up relating them.

[6] See Ernst Tugendhat, 'La pretensión absoluta de la moral y la experiencia histórica' (1979) and Chapter 8 of this book.

[7] Ernst Mayr contrasts functionalist and evolutionary demands in biology, although he considers that they are concerned with complementary aspects of what would be a complex phenomenon in need of both types of approach ('Cause and Effect in Biology' 1961). Functionalist interpretations of capabilities such as rationality and moral sense are obviously of great importance in supporting the hypothesis of natural selection. The relation between these different aspects can be found in recent articles such as those of Elliott Sober, 'The evolution of rationality' (1981) and Robert N. Brandon, 'Biological Teleology: Questions and Explanations' (1981), while the possible meanings of a functionalist explanation in biology have been discussed by Ernest Nagel in the second part of his article 'Teleology revisited' (1977).

Of course, the functionalist theory of morals as a significant element for evolution and selection is neither universally accepted nor foreign to criticism. Hilary Putnam in 'Why reason can't be naturalized' (1982) in fact attacks any argument of this type, denying the viability of attempts at evolutionary episte-mology of the type used in the Darwinian naturalization of human rationality. If rationality is the capacity to arrive at beliefs which promote our survival, that is, our inclusive fitness (something which is taken for granted in this book) we are squarely placed in a line of argument which leads, according to Putnam, to the need to speculate a metaphysical notion of 'truth'. I think Putnam's attempt is correct as a critique of naturalization of the process of rational preference, but I also consider that not *all* propositions which are supported by the Darwinian inheritance and which at the same time revindicate the role of reason in such matters must necessarily stumble over this pitfall. Later on I shall deal with the objections of Kantian metaphysics to the link between nature and reason and the need for a base of causal reasoning to sustain a model of rational preference compatible with the functionalist theory of ethics from the biological perspec-tive. I consider that in this way Putnam's critique can be avoided and we can continue along the line which will take us to the propositions contained in Chapter 8.

[8] Between the works of Hume and Smith and that of Darwin there is not, of course, just empty space. Authors such as Thomas Reid, Dugald Stewart, Thomas Brown, and Mackintosh himself continue the critiques of the relations between moral sense and ethical criterion, and in these the solutions offered by Hume and Smith are analysed. Of special interest in this context is a work as accessible to Darwin as that of Alexander Bain. If only Hume and Smith are dealt with here it is because in my opinion the operations which modify the classic theory of moral sense in these authors represent the key to interpreting relations between the alpha and beta-moral levels. For Smith's influence on Reid's work see, in addition to the article quoted by Macintyre, the polemic between Elmer H. Duncan and Robert M. Baird (1977) and David F. Norton and J. C. Stewart-Robertson (1980). Hume's influences on Darwin's work have been analysed in a general sense, somewhat removed from the subject dealt with here, by William B. Huntley (1972). For the question of the role played by sympathy in the moral sense school, see C. J. Cela Conde and Alberto Saoner (1979); I have liberally used material from that article here, although Professor Saoner cannot be held responsible for what I state now.

[9] See note 8 above.

[10] Reference is made here principally to the contents of Section V, part II, paragraphs 576 ff. of D. D. Raphael's edition, *British Moralists*.

[11] See further on, Chapter 5.

[12] This is a central idea in neo-Darwinism and appears as much in Julian Huxley, *Evolution and Ethics* (1947) as in Dobzhansky, *The Biological Basis of Human Freedom* (1956) and Waddington, *The Ethical Animal* (1960).

[13] Cf. for example, D. D. Raphael, *Darwinism and Ethics* (1958) and Mar-jorie Grene, *The Ethical Animal: A review* (1962). Cultural functionalism in the theory of evolution has been criticized by B. H. Williams (1980). (See note 4 to this chapter).

CHAPTER 3

[1] I will return to this matter in the following chapter.

[2] *Critique of Pure Reason.*

[3] Cf. Kolakowski's critique of this interpretation in *Husserl and the Search for Certitude* (1975), pp. 20–30, and chapter 4 of this book.

[4] Those who would like a bibliography should see my article 'Una aproximación a la 'hipótesis de las ideas innatas' de Noam Chomsky' (1976). I must confess that my current position on this subject does not entirely coincide with what I wrote there.

[5] Fundamentally because he did not wish his hypothesis on innate ideas to be confused with innatisms postulated by ethologists which have been systematically used to support extreme right-wing political standpoints.

[6] The innatist model of cognitive psychology will not be taken into account here, among other reasons because it is reducible at least in part to Chomskyan theses which I use as a point of departure which is quite adequate, in my opinion. Cf. on relations between cognitive psychology and ethical matters, M. Jiménez Redondo 'Teorías contemporáneas del desarrollo moral' (1979), and Hugh Rosen *The Development of Sociomoral Knowledge* (1980). The latter includes Chomsky and Levi-Strauss in the same category of 'modern structuralists', whom he accuses of staying on the sidelines of genetics, and consequently leans towards Piaget and Kohlberg in most of his work. I imagine it is not necessary to emphasize that this opinion seems to me to be absurd, above all because it is voiced after the appearance of Lenneberg's book (1967), which is neither quoted in the pages of the text nor in Rosen's bibliography. See the end of this chapter.

[7] Wilson had already proposed the programme of a CST in an article entitled 'Comparative Social Theory' (1980).

[8] In 'Sociobiology and Human Nature: a postpanglossian vision' (1980) Stephen Jay Gould has undertaken a critique of the inferences made by Wilson in *On Human Nature* about the relations existing between culture and genetic programming, choosing a more classical alternative within the lines of the 'Promethean gene' which comes from Dobzhansky. After Lumsden and Wilson it does not seem very effective to use arguments of the type Gould employs, which refer, for example, to the role of culture as an alternative to genetic content in the explanation of certain adaptive features. Given that Wilson and Lumsden use (as the former did earlier) a concept of culture which in itself includes genetic determinism, the typically 'culturalist' arguments are out of place. I think criticisms of Wilson should in principle accept such a foundation as a strategic hypothesis and analyse its possible shortcomings. At least that is the line which will be followed here.

[9] Rawls moved from linguistic competence to ethical competence in a way which is certainly related to what I am trying to establish from Chomskyan postulates about innate ideas. Such an idea has been criticized by Thomas Nagel (1973, p.2) for whom the analogy was false, since intuitions of the native speaker in fact form language, ethics is not made up of analogous intuitions and, in any case, the plausibility of an ethical theory can cause a change in our moral

intuitions. This specification of Nagel is absolutely correct as long as we are
referring to intuitions relative to the gamma content of empirical codes. As I
trust will become clear later, moral intuition which can rest on biological
capabilities, and make use therefore of the Chomskyan model of competence,
refers to beta-moral aspects, that is to structural contents of ethical reference
linked to the sense of an evaluative word like 'right' and not to empirical
assignations of the evaluation made within a given community. I must point out
that, in any case, linguistic behaviour is not free of empirical changes either,
changes which convert a language into something extremely fluid and unstable.

CHAPTER 4

[1] A distribution however which also fails to solve the problem. One of the
authors who has studied Moore's arguments on this matter in the greatest detail,
Stephen Toulmin, classifies values ('good') either as a class apart from properties ('yellow') or as a special, 'non-natural', class of properties (1960, chapter II;
Toulmin declines however to use a classification which is epistemologically
superficial).

[2] At times Ayer has maintained a more orthodox 'analytic' view, such as
when he indicates the existence of four types of ethics which are commonly
confused in the works of philosophers: (a) propositions which express defini-
tions of ethical terms or judgements about their legitimacy or possibility; (b)
propositions which describe phenomena of moral experience and its causes;
(c) exhortations to moral virtue, and (d) truly ethical judgements. For Ayer only
the first of the four constitutes a moral philosophy *stricto sensu*. With regard to
the second, he postulates its assignation to the science of psychology or sociology. The third and fourth are rejected as foreign to philosophy and/or to science
(1967, pp. 119 ff).

[3] Compare for example the successive *readings* edited by M. F. Ashley
Montagu devoted to the ethology of Lorenz (1968) and the sociobiology of
Wilson (1980).

[4] The literature on this is immense as it is one of the central theses of
sociobiology. By way of example, consult the bibliography which follows the
study by Danilo Mainardi (1980) of the evolving convergence which affects the
capacity to transmit habits and social information, and the empirical evidence
existing about this. Needless to say, *Sociobiology* by Wilson also contains ample
information on this subject.

[5] See Eibl-Eibesfeldt (1967) Chapter 18, paragraph 5. On innate programmes
of the submission to and acceptance of hierarchies, apart from the contributions
of ethologists, see J. Z. Young *Programs of the Brain* pp. 222 ff.

[6] Genetic drift is a change in the frequency of genes in a population owing to
factors of chance and the intervention of the phenomenon known as 'sampling
error'. For these reasons, in relatively small populations the genetic frequency
may be different from that which would theoretically be attributable through the

strict application of parameters defined by population genetics. (See Ayala and Valentine 1979, pp. 105 ff; Wilson, 1975, pp. 66 ff). The presence of genetic drift can give rise to different evolution from that of natural selection, with neutral genes (those without selective value) establishing themselves. See Wilson and Bossert (1971), pp. 88 ff.

[7] In the sense used by Eibl-Eibesfeldt (See note 5). As early as 1929 David Ross speculated about the original use of the word 'good' in bygone times, posing the possibility thereafter of an interjection or expression of admiration in the context of the analysis of the different meanings of the word (1939, p. 244). Peter Singer also uses the phylogenetic analysis of social practices to evaluate the functional role of reason and morality (1981, Chapter 4, pp. 90 ff. under the title 'The first step').

[8] 1967, pp. 525 ff.

[9] The presence of mental structures capable of including ethical universals can also refer to proposals different from Stroll's as far as the acquisition of language is concerned: for example, to Thomas Nagel, 'What is it like to be a bat?', (Chapter 12 of *Mortal Question*, 1979) or to Viki McCabe, 'The direct perception of universals: A theory of knowledge acquisition' (1982). I am not going to enter into this debate of contemporary philosophy of the mind; I should just like to point out that there are more than enough speculative foundations upon which to build a theory of innate beta-moral structures. But, now we are on the subject, I shall comment on one aspect of the last article mentioned above.

McCabe proposes an alternative to the classic empirical and rationalist theories about relationships between universals and knowledge, maintaining that universals are schemes existing in the real world and directly perceived in the task of knowledge (we perceive the 'schemes' of idealists directly from the 'external world' of realists, rather than the 'components' of realists in the 'internal world' of idealists). McCabe locates the reason for holding this opinion in the biological value of the perception of structural and transformational invariants, something with which I agree. I do not agree so much with the explanation given by McCabe about the cognitive differences which can exist between species, supposedly due to the perception of different parts of the world and not to different structural properties of the mind. I think McCabe's article, although of course very thought provoking, meets with two types of difficulty: (1) it underrates the human mind's character of 'thing in the world', with its own structural conditions, and (2) he proposes a model of knowledge which is not very different from those he classifies under the title of 'empirical idealism', such as proposed by J. J. Gibson (1979) and in the final analysis, by Hume himself. There is not a great difference between proposing that different species form a different epistemological idea of the world, and claiming that the world remains constant and each species perceives a part (not a translation) of it; or at least there is not a great difference to the biological value of under-standing the world. On the contrary, serious difficulties arise if we claim to situate certain characteristics of representation exclusively in the world such as all those referring to universal ethics, to which I am making continual reference.

Such invariants are, in my opinion, properties of a typical type of knowledge that is human, and are neither the invariant properties of external schemes or features (inasmuch as we should consider as 'internal' the semantic aspect of language), nor invariant properties of other species' means of representing the world (and in this sense Wilson's attempt to compare human ethics to the ethics of termites is mistaken.)

[10] The concept of 'evolutionary stable strategy' comes from the application of games theory to evolution, in order to achieve formal models of prediction as a result of maintaining certain lines of conduct. It was J. Maynard Smith who proposed this concept, even though, according to him, the original idea of applying mathematics to the evolution of conflicts between animals was George Price's (See the preface to *On Evolution* 1972). A 'stable evolutionary strategy' is a line of conduct which cannot be surpassed, in terms of adaptation, by another alternative line, providing that the first is followed by the majority of individuals in a population. For more details see J. Maynard Smith and G. R. Price 'The Logic of Animal Conflicts' (1973); J. Maynard Smith, 'Evolution and the Theory of Games' (1976); R. C. Lewontin, 'Evolution and the Theory of Games' (1981); D. McFarland and A. Houston, *Quantitative Ethology* (1981) (above all Chapters 6–9), and Richard Dawkin's synthesis in Chapter 5 of *The Selfish Gene* (1976). A warning about the risks arising from a rigid application of the Theory of Games is to be found in James Silverberg (1980) pp. 47–49.

[11] The distinction between 'open programmes of conduct' (in the sense that a line of conduct, genetically fixed, gives rise to different options among which the individual must choose) and 'closed programmes of conduct', which are rigidly controlled by the genetic code) comes from Ernst Mayr (1976). It was Popper, however, who did most to popularize the concept by relating it to processes of emergence to illustrate the adaptive value of the language/rational thought binomial. The debt which the theses I maintain throughout this book owe to those of Popper (expressed for example in 'Natural Selection and the Emergence of Mind' 1978, or the book written jointly with Eccles, *The Self and its Brain*, 1977) is I hope sufficiently clear. Although we are only dealing with moral behaviour here, the weight that linguistic expression and reason carry in such matters is difficult to ignore. A good exposition of the significance of Popperian theses and their role in the field of thought related to the evolution of the capacity to think and know can be found in G. Radnitzky, 'The science of man: Biological and cultural evolution' (1983). I must point out however that Popper's aim goes well beyond the level referring to the adaptive capacities of human language to the metaphysical interpretation of the famous 'three worlds' a subject outside my current interest. On the heuristic value of the Popperian 'world 3' and its ontological ambiguity, consult Gregory Currie's 'Popper's evolutionary epistemology: a critique' (1978. A more favourable view of Popper's theses can be found in the article mentioned by Radnitzky. Compare also D. T. Campbell 'Evolutionary Epistemology' (1974), and Popper's response contained in the second volume for the effects of the Popperian stance on the evolution of knowledge.

[12] Francisco J. Ayala considers that teleological explanations can only be

maintained within biology as a science, therefore making it irreducible to physics or chemistry. (See Section III 'The notion of teleology' in 'Biology as an Autonomous Science' 1968; compare **Marjorie Grene's** critique of this idea in 'Explanation and Evolution 1974). The expression 'la logique du vivant' obviously refers to the well known book by François Jacob, and the emphasis on the functions of invariance and teleonomy no less evidently to the theses of Jacques Monod.

[13] That is the essence of the critique made in Chapter 3 of the proposals by Lumsden and Wilson contained in *Genes Mind and Culture*.

[14] The topic of freedom and will in Kant are set out in Leslie W. Beck *A commentary on Kant's Critique of Practical Reason*, Chapter XI, 'Freedom' (1960) and Kant's two conceptions of the will in their political context' (1965).

[15] The releasing role of moral language has been made more than clear enough by Stevenson through his analysis of the emotive elements in discourse, and especially within the study of persuasive methods (*Ethics and Language* 1944, Chapter VI).

[16] See the critique of Alexander's theses about the concept of 'goodness' made in Chapter 1.

CHAPTER 5

[1] See Javier Muguerza, *La razón sin esperanza* (1977), chapter VII, which will be returned to later on.

[2] See Chapter 2.

[3] The synthesis by W. D. Hudson, *Modern Moral Philosophy* (1970) Chapter 5, part III, paragraph II can be consulted.

[4] The principle of axiological neutrality (*Wertfreiheit*) invades quite early on, from Weber, the field of discussion about human sciences, but to a certain extent the impression exists that this vintage has only added force to the controversy about the relevance of such affectations if we take account of the prolonged duel between the neopositivists and dialecticians on this point. Even so, it appears that in recent years the controversy has lost a certain force confronted with the proposals of synthesis made by the Erlangen school (such as the reconstruction of practical reason by Lorenzen).

[5] The literature triggered by the appearance of Rawl's neocontractualism from his first articles in the 1950's is a little too overwhelming to make precise references. I should like to draw special attention to R. M. Hare's critique, 'Rawls' Theory of Justice' (1973), the articles by J. H. Schaar, 'Reflections on Rawls' Theory of Justice' (1974) Thomas Nagel, 'Rawls on Justice' (1973) and Michael Gorr, 'Rawls on natural inequality' (1983), and the books by R. Nozick, *Anarchy, State and Utopia* (especially Chapter 7, Section II) (1974), B. Barry, *The Liberal Theory of Justice* (1974), Robert Wolff, *Understanding Rawls* (1977) and D. D. Raphael, *Justice and Liberty* (especially Chapter 7) (1980).

[6] On etic *versus* emic rationality, see Chapter 7 of this book.

[7] Formalism as a school has turned out to be especially fertile, apart from economic theory (but in close relation to it) within the field of anthropology. See for example R. Burling, 'Maximization Theories and the study of Economic Anthropology' (1962), E. E. Le Clair, 'Economic Theory and Economic Anthropology' (1962), and in general the anthology by M. Godelier, *Un domaine contesté: l'antropologie économique* (1974). The ringleader on the substantive front opposing formalism is Polanyi whose 'The Economy as Instituted Process' (1975) can be consulted.

[8] I should like to thank Victoria Camps, who presented a paper entitled 'Etica y retórica' at the *Primeras Jornadas de Etica e Historia de la Etica* held in the UNED of Madrid in 1979, for calling my attention to Perelman's work and lending me a copy of her paper.

[9] In the somewhat ingenuous assumption that that would be the purpose of parliamentary speeches.

CHAPTER 6

[1] The central dogma of molecular genetics in the formulation used by Monod does not seem free of all suspicion for reasons which are both empirical (the possibility of backward movement of information from the protein to RNA) and theoretical (like those adduced by René Thom, 1968, p. 61 when he refers to the reciprocal effect of the receptor on the source within the theory of information). The idea that I use regarding the environment as a source of channelling of stochastic process has little to do with René Thom's objection and is commonplace within the 'orthodox' theses of molecular biology.

[2] This is one of the foundations for Marvin Harris' theory of cultural materialism. See for example *Cannibals and Kings* (1977).

[3] The reader interested in the mathematical justification of the model can consult E. Wilson, *Sociobiology* (1975, pp. 83 ff.).

[4] See John H. Crook, *The Evolution of Human Consciousness* (1980 pp. 184 ff.)

[5] Empirical experience shows us that birth rate decreases and mortality increases as the population grows and per capita resources are in consequence reduced. To adjust the shape of the curve expressing this variation could be a somewhat tedious task, but in an initial analysis one could assume that the variations in birth rate b and mortality d are both linear in relation to the value of the total population N. This is probably too crude an assumption since it is easy to understand that the availability of resources is not going to decrease in a linear way compared to the population which is growing. Rather it seems that the first increases would not make much difference in relation to the initial values of birth and death rates (at least by the isolated intervention of this factor), while, as resources in the environment approach their limits, great demographic disturbances are to be expected. This forces one to think of an exponential shape to the curve which represents the variation of b and d as a function of N, but to simplify the model we can keep to the idea of linearity.

[6] See F. Rodríguez Adrados, *La democracia ateniense* (1975), Chapter 4.

[7] A complete account of the investigations inspired by Wittfogel's theory of hydraulic despotism can be found in J. R. Llobera, 'Karl Wittfogel y el modo asiático de producción' (1980).

[8] *Essay concerning the True, Original, Extent and End of Civil-Government* (second of the *Two Treatises of Government*, (1698) paragraph 28, pp, 20–30).

[9] To which is no reason to concede bill of universality.

CHAPTER 7

[1] Mary Midgley completely rejects Wilson's arguments on the philosophical value which considerations about the hypothalamic-limbric system could have (not mentioned in the quote included here) suggesting that mathematics is also a significant part of human thinking regarding evolution and nobody takes it into their head to explain mathematics and mathematicians by dissecting a brain (1979, pp. 169 ff). Midgley's rejoinder points to a somewhat weak flank to Wilson's theses, but even assuming a certain argumentative insecurity on his part in the philosophical field, the body of proposals about ultimate ends which have been repeatedly stressed in the Darwinist tradition still holds.

[2] The distinction is owed to Kenneth Pike (1954) and has been systematized by Marvin Harris among others.

[3] Jesús Mosterín proposes the following concept of practical rationality: 'We say that individual x is rational in his conduct if (1) x is clearly aware of his ends, (2) x knows (as far as possible) the means necessary to achieve these ends, (3) if as far as possible x puts adequate means into action to achieve these ends, (4) in case of conflict between ends of the same type and of different degrees of proximity, x gives preference to the latter, and (5) the ultimate ends of x are mutually compatible' *Racionalidad y acción humana* (1978, p. 30). Being far more detailed than my own, and without doubt more interesting, Mosterín's definition to my way of thinking suffers from certain ambiguities ('as far as possible') which diminish its claim to rigour. I think it is better to refer to 'common sense' contained in the practices of tradition. Cf. Javier Muguerza (1977) pp. 194–195.

[4] Cf. Javier Muguerza (1977) pp. 212 ff.

[5] Quoted by J. L. Aranguren, *Etica* (1958), p. 96.

[6] In 'Darwinism and Ethics' (1958). Waddington replies in *The Ethical Animal* (1960) (rather inadequately in my view) and Marjorie Grene joins in the dispute stressing Raphael's arguments ('The Ethical Animal: A review,' 1962).

CHAPTER 8

[1] The idea of fairness as a structural property of language has also been dealt with by Mary Midgley (1978), p. 225 ff.

[2] See Robert Richards (1982) p. 53.

[3] In *Animal Behaviour in Relation to Social Behaviour* (1962). To the two classic models of group and kin selection others have recently been added which in my judgement do not greatly modify the panorama presented here, such as Trivers' model of reciprocal altruism (1971) or Wilson's model of synergistic selection (1975). *Cf.* J. Maynard Smith, 'The evolution of social behaviour – a classification of models' (1982).

[4] See Hamilton (1964). The biological concept of *fitness* has however a long history prior to the sense given by Hamilton (see R. I. M. Dunbar, 1982, pp. 13–14). See also Chapter 1 of this book, note 5.

[5] See Chapter 1, note 3 and Chapter 4, note 10. A formalization of the model of kin selection with alternative calculations of the increase in inclusive fitness has been synthesized by Peter O'Donald, 'The concept of fitness in population genetics and sociobiology' (1982).

[6] Peter Singer indicates that reciprocal altruism, kin altruism and a limited amount of group altruism may all have been developed among the social animals from which we descended (1981, p. 54) and suggests later on (p. 91) that in a small community, tendencies towards group altruism (which supplements the stronger tendency towards kinship altruism) can appear as a line of conduct in preference to altruism towards human beings belonging to other different groups.

CHAPTER 9

[1] Classic models of this type were of course abundant in the last century with the rise of innatism. Those of Cesare Lombroso *The Man of Genius* (1891) and Sir Francis Galton, Darwin's cousin *Hereditary Genius* (1892) are well known. Today the controversy runs along somewhat different channels which do not bear too close a relation to what I am trying to argue here (such as Eysenck's theses). The interested reader can turn in any case either to the Freudian psychoanalytic tradition or to the acts of symposia such as the one held in Toronto in (1969) (W. B. Dockrell, ed. *On Intelligence* 1970), or the 7th Hyman Blumberg symposium (J. C. Stanley, W. C. George and C. H. Solano, eds., *The Gifted and the Creative: A fifty-year perspective*); etc. See also W. Dennis and M. W. Dennis (eds), *The Intellectually Gifted* (1976) and Pieter A. Vroon, *Intelligence. On Myths and Measurement* (1980). In my view the most interesting perspective is that established by John Hartung when he considers the relations between natural selection and the inheritance of wealth, understanding by wealth a wide set of resources, talents and status able to increase the reproductive success of whoever possesses them ('On Natural Selection and the Inheritance of Wealth' 1976, p. 607). Hartung, especially in the explanation he gives to critics of his article, presents an idea of the interaction between cultural norms, inherited conducts and natural selection which was then widely discussed in the sociobiological context and not always as clearly as in Hartung (op. cit. p. 619) See also J. Hartung, 'Paternity and the inheritance of wealth' (1981).

[2] See Francisco J. Ayala, 'The Concept of Biological Progress' (1974) on the criteria which permit progress in biology to be defined.

[3] See the concept of autotrophism in Faustino Cordón (1981) Chapter IV.

[4] The interested reader can refer to my article 'Tres tesis falaces de la ideología liberal' (1982).

[5] Although it had already been pointed out by Konrad Lorenz (1963) Chapter XIII.

[6] Hayek uses the sentence by hypothetically, and goes on to deny the hypothesis.

[7] *Law, Legislation and Liberty*, Vol. III (1979), p. 164.

BIBLIOGRAPHY

Carlos R. Alba Tercedor y Fernando Vallespín (1979): "El neocontractualismo de 'A Theory of Justice' de John Rawls, una introducción a la literatura", in *Revista de Estudios Políticos*, **8**, pp. 233–250.

Richard D. Alexander (1977): *Darwinism and Human Affairs*. London, Pitman.

José Luis L. Aranguren (1958): *Etica*. M., Revista de Occidente. Quoted from the 6th ed., 1976.

John Austin (1832): *The Province of Jurisprudence Determined*. London, John Murray (2nd ed., extended, 1861–1863, 3 vols.).

Francisco J. Ayala (1968): 'Biology as an Autonomous Science', in *American Scientist*, **56**, pp. 207–221. Quoted from the ed. contained in M. Grene and E. Mendelsohn (eds.), *Topics in the Philosophy of Biology* (1976): pp. 312–329.

Francisco J. Ayala (1974): 'The Concept of Biological Progress', in F. J. Ayala and T. Dobzhansky (eds.), *Studies in the Philosophy of Biology*, pp. 339-355.

Francisco J. Ayala (1980): *Origen y evolución del hombre*. M., Alianza Editorial.

Francisco J. Ayala and Theodosius Dobzhansky (eds.) (1974): *Studies in the Philosophy of Biology*. London, Macmillan.

Francisco J. Ayala and J. W. Valentine (1979): *Evolving. The Theory and Processes of Organic Evolution*. Menlo Park, Cal., Benjamin Cummings. Quoted from the Spanish ed, *La evolución en acción*, trans. by M. Ochando, M., Alhambra, 1983.

Alfred J. Ayer (1967): *Language, Truth and Logic*. London, Gollancz. Quoted from the Spanish ed., *Lenguaje, verdad y lógica*, trans. by M. Suárez, B., Martínez Roca, 1971.

Samuel Bailey (1855–1863): *Letters on the Philosophy of the Human Mind. Third Series*. London, Longman, Brown, etc.

Alexander Bain (1868): *Mental and Moral Science. A Compendium of Psychology and Ethics*. (2nd ed., 1872). London, Longman.

George W. Barlow and James Silverberg (eds.) (1980): *Sociobiology: Beyond Nature/Nurture?* Boulder, Col., Westview Press.

S. A. Barnett (ed.) (1958): *A Century of Darwin*. London, Heinemann. Quoted from the Spanish ed, *Un siglo después de Darwin*, trans. by F. Cordón, M. Alianza Editorial, 1966 (2 vols.).

Brian Barry (1974): *The Liberal Theory of Justice*. Oxford, Clarendon Press.

Leslie White Beck (1960): *A Commentary on Kant's Critique of Practical Reason*. University of Chicago Press.

Leslie White Beck (1965): 'Kant's two conceptions of the will in their political context'. Chapt. XIII of *Studies in the Philosophy of Kant*, N.Y., Bobbs Merrill. A first French ed. of this article, 'Les deux concepts Kantiens du vouloir dans leur contexte politique', appeared in *Annales de philosophie politique* **VI**, 1962, pp. 119–137.

Brian C. R. Bertram (1982): 'Problems with altruism', in King's College

Sociobiology Group (ed.), *Current Problems in Sociobiology*, pp. 251–267.

Robert N. Brandon (1981): 'Biological Teleology: Questions and Explanations', in *Studies of History and Philosophy of Science*, **12**, pp. 91–105.

C. D. Broad (1944–1945): 'Some Reflections on Moral-Sense Theories in Ethics', in *Proceedings of the Aristotelian Society*, **45**.

Thomas Brown (1820): *Lectures on the Philosophy of the Human Mind*. Edinburgh, D. Welsh (4 vols.).

Robbins Burling (1962): 'Maximization Theories and the Study of Economic Anthropology', in *American Anthropologist*, **64**, pp. 802–821.

Donald T. Campbell, 'Evolutionary Epistemology', in Paul A. Schilpp (ed.), *The Philosophy of Karl Popper*, La Salle, Open Court, 1974, pp. 413–463.

Victoria Camps (1979): 'Etica y retórica'. *Actas de las primeras jornadas de Etica e Ha. de la ética*, M., U.N.E.D. (in press).

V. Cappelletti, B. Luiselli, G. Radnitzky and E. Urbani (eds.) (1983): *Saggi di storia del pensiero scientifico*. Rome, Jouvence.

Camilo J. Cela-Conde (1976): 'Una aproximación a la 'Hipótesis de las ideas innatas' de Noam Chomsky', in *Mayurqa*, **15**, pp. 139–188.

Camilo J. Cela-Conde (1982): 'Tres tesis falaces de la ideología liberal', in *Sistema*, **50-51**, pp. 51–60.

Camilo J. Cela-Conde and Alberto Saoner (1979): 'En torno al concepto de simpatía. Un capítulo en el desarrollo del pensamiento liberal'. *Actas de las primeras jornadas de Etica e Ha. de la ética*, M., U.N.E.D. (in press).

François Chatelet (dir.) (1978): *Histoire des Idéologies*. Paris, Hachette.

Michel Clévenot (1976): *Approches matérialistes de la Bible*. Paris, Cerf.

Noam Chomsky (1980): *Rules and Representations*. Oxford, Basil Blackwell.

Faustino Cordón (1981): *La naturaleza del hombre a la luz de su origen biológico*. B., Anthropos.

John H. Crook, *The Evolution of Human Consciousness*, Oxford, Clarendon Press, 1980.

Gregory Currie (1978): 'Popper's evolutionary epistemology: A critique', in *Synthese*, **37**, pp. 413–431.

Charles R. Darwin (1871): *The Descent of Man and Selection in Relation to Sex*. (2nd ed., modified, 1874). Quoted from the N.Y. Modern Library ed., 1936 (numerous impressions) which follows the 2nd ed. of 1874, and also contains the *Origin of Species*.

Richard Dawkins (1976): *The Selfish Gene*. Oxford University Press. Quoted from the Spanish ed., *El gen egoísta*, trans. by J. Robles, B., Labor, 1979.

Wayne Dennis and Margaret W. Dennis (eds.) (1976): *The Intellectually Gifted* N.Y., Grune & Stratton.

Theodosius Dobzhansky (1956): *The Biological Basis of Human Freedom*. Columbia University Press.

W. B. Dockrell (ed.) (1970): *On Intelligence*. London, Methuen & Co.

R. I. M. Dunbar (1982): 'Adaptation, fitness and the evolutionary tautology'. In King's College Sociobiology Group (ed.), *Current Problems in Sociobiology*, pp. 9–28.

Elmer H. Duncam and Robert M. Baird (1977): 'Thomas Reid on Adam

Smith's Theory of Morals', in *Journal of History of Ideas*, **38**, pp. 509–522.

Irenäus Eibl-Eibesfeldt (1967): *Grundiss der vergleichenden Verhalstenforschung*. Munich, R. Piper. Quoted from the Spanish ed., *Etología*, trans. by M. Costa, B., Omega, 1974.

Friedrich Engels (1878): *Anti-Dühring*. Quoted from the Spanish ed. of same title, trans. by J. Verdes, M., Ciencia Nueva, 1968.

E. E. Evans-Pritchard (1937): *Witchcraft, Oracles and Magic among the Azande* Oxford, Clarendon Press. Quoted from the Spanish ed., *Brujería, magia y oráculos entre los Azande*, trans., by A. Desmonds, B., Anagrama 1976.

José Ferrater Mora and Priscilla Cohn (1981): *Etica aplicada*. M., Alianza Editorial.

Paul K. Feyerabend (1978): *Science in a Free Society*. London, New Left. Quoted from the 2nd ed., 1962.

R. Firth (1952): 'Ethical Absolutism and the Ideal Observer', in *Philosophy and Phenomenological Research*, **12**.

Philippa Foot (1978): *Virtues and Vices*. Oxford, Basil Blackwell.

Francis Galton (1892): *Hereditary Genius*. London, Macmillan.

J. J. Gibson (1979): *The Ecological Approach to Visual Perception*. Boston, Houghton Mifflin.

P. T. Geach (1977): *The Virtues*. Cambridge University Press.

Maurice Godelier (ed.) (1974): *Une domaine contesté: l'anthropologie économique*. Paris, Mouton.

Michael Gorr (1983): 'Rawls on natural inequality', in *The Philosophical Quarterly*, **33**, pp. 1–18.

Stephen Jay Gould (1980 a): 'Sociobiology and Human Nature: a postpanglossian vision', in A. Montagu (ed.), *Sociobiology Examined*, pp. 283–290.

Stephen Jay Gould (1980 b): *The Panda's Thumb*. Harmondsworth, Penguin.

Marjorie Grene (1962): '*The Ethical Animal*: a review', in *British Journal of the Philosophy of Science*, **13**, pp. 173–176. Quoted from the ed. included in M. Grene's *The Understanding of Nature* (1974).

Marjorie Grene (1974): 'Explanation and Evolution'. Chapter XIII of *The Understanding of Nature*, Dordrecht, Reidel.

Marjorie Grene and Everett Mendelsohn (eds.) (1976): *Topics in the Philosophy of Biology*. Dordrecht, Reidel.

Irving Hallowell (1960): *Evolution After Darwin*. University of Chicago Press.

W. D. Hamilton (1964): 'The Genetical Theory of Social Behaviour', in *Journal of Theoretical Biology*, **7**, pp. 1–16 and 17–32.

R. M. Hare (1963): *Freedom and Reason*. Oxford University Press.

R. M. Hare (1973): 'Rawls' *A Theory of Justice*' in *Philosophical Quarterly*, **23**, pp. 144–155 and 241–252.

Marvin Harris (1977): *Cannibals and Kings: The Origin of Cultures*. N.Y., Random House.

John Hartung (1976): 'On Natural selection and the inheritance of wealth', in *Current Anthropology*, **17**, pp. 607–622.

John Hartung (1981): 'Paternity and inheritance of wealth', in *Nature*, **291**, pp. 652–654.

F. A. Hayek (1979): *Law, Legislation and Liberty*. Vol. III: *The Political Order of a Free People*. London, Routledge and Kegan Paul.

Martin Heidegger (1929): *Was ist Metaphysik?* Separate ed. of *Jahrbuch für Philosophie und Phenomenologische Förschung*, **10**. Spanish ed. in *Cruz y Raya*, **6**, 1931.

Douglas R. Hofstadter (1983): 'Temas metamágicos', In *Investigación y ciencia* (Spanish ed. of *Scientific American*), **83**, pp. 102–107.

W. D. Hudson (1970): *Modern Moral Philosophy*. London, Macmillan. Quoted from the Spanish ed., *La filosofía moral contemporánea*, trans. by J. Hierro, M., Alianza Editorial, 1974.

David Hume (1793–1740): *A Treatise of Human Nature*. Ed. by L. A. Selby-Bigge (2nd ed., by P. H. Nidditch), Oxford University Press, 1978.

David Hume (1751): *An Enquiry concerning to the Principals of Morals*. (Revised ed. in 1777). In D. D. Raphael (ed.), *British Moralists* (1969), vol. II.

William B. Huntley (1972): 'David Hume and Charles Darwin', in *Journal of the History of Ideas*, **32**, pp. 457–470.

Julian S. Huxley (1947): *Evolution and Ethics*. London, Pilot Press.

Uffe J. Jensen and Rom Harré (eds.) (1981): *The Philosophy of Evolution*. N.Y., St. Martin's Press.

Manuel Jiménez Redondo (1979): 'Teorías contemporáneas del desarrollo moral', in José Rubio Carracedo, Manuel Jiménez Redondo and Jesús Rodríguez Marín, *Génesis y desarrollo de lo moral*, pp. 63–136.

Immanuel Kant (1781): *Kritik der reinen Vernunft*. Spanish ed. by Pedro Ribas, *Crítica de la razón pura*, M., Alfaguara, 1978.

King's College Sociobiology Group (ed.). *Current Problems in Sociobiology*. Cambridge University Press, 1982.

Leszek Kolakowski (1975): *Husserl and the Search for Certitude*. Yale University Press. Quoted from the Spanish ed., *Husserl y la busqueda de certeza*, trans. by A. Murguía, M., Alianza Editorial, 1977.

Edward E. LeClair Jr. (1962): 'Economic Theory and Economic Anthropology', in *American Anthropologist*, **64**, pp. 1, 179–1, 203.

Eric H. Lenneberg (1967). *Biological Foundations of Language*. N.Y., Wiley. Quoted from the Spanish ed., *Fundamentos biológicos del lenguaje*, trans. by N. Sánchez and A. Montesinos, M., Alianza Editorial, 1975.

R. C. Lewontin (1976): 'Evolution and the Theory of Games', in M. Grene and E. Mendelsohn (eds.), *Topics in the Philosophy of Biology*, pp. 286–311.

Philip Lieberman (1973): "On the evolution of language: A unified view", in *Cognition*, **2**, pp. 59–94. Quoted from the Spanish ed. "Un enfoque unitario de la evolución del lenguaje". in V. Sánchez de Zavala (ed.), *Sobre el lenguaje de los antropoides* (1976), pp. 147–203.

Josep R. Llobera (1980): 'Karl Wittfogel y el modo de producción asiático'. In *Hacia una historia de las ciencias sociales*, B., Anagrama, capítulo V.

John Locke (1690–1698): *Two Treatises of Government*. Ed. by Peter Laslett, Cambridge University Press, 1960.

Cesare Lombroso (1891): *The Man of Genius*. London, Walter Scott.

Konrad Lorenz (1963): *Das sogenante Böse*. Viena, Borotha-Schoeler. Quoted from the Spanish ed., *Sobre la agresión*, trans. by F. Blanco, M., Siglo XXI, 1971.

Charles J. Lumsden and Edward O. Wilson (1981). *Genes, Mind and Culture. The Coevolutionary Process*. Harvard University Press.

Alasdair Macintyre (1982). 'How moral agents became ghosts, or why the History of Ethics diverged from that of the Philosophy of Mind', in *Synthese*, **53**, pp. 295–312.

James Mackintosh (1836). *Dissertation on the Progress of Ethical Philosophy*. Edinburgh, Adam & Charles Black.

Danilo Mainardi (1980). 'Tradition and the Social Transmission of Behaviour in Animals', in G. W. Barlow and J. Silverberg (eds.), *Sociobiology: Beyond Nature/Nurture?*, pp. 227–255.

Edward Manier (1978). *The Young Darwin and his Cultural Circle*. Dordrecht, Reidel.

Ernst Mayr (1961): 'Cause and Effect in Biology', in *Science*, **134**, pp. 1, 501–1, 506.

Ernst Mayr (1976): *Evolution and the Diversity of Life*. Harvard University Press.

Viki McCabe (1980). 'The direct perception of universals: A theory of knowledge acquisition', in *Synthese*, **25**, pp. 465–513.

D. MacFarland and A. Houston (1981). *Quantitative Ethology*. Boston, Mass., Pitman.

Sterling McMurrin (ed.) (1980). *The Tanner Lectures on Human Values*. Cambridge University Press.

Richard E. Michod (1981). 'Positive Heuristics in Evolutionary Biology'. in *The British Journal of the Philosophy of Science*, **32**, pp. 1–36.

Mary Midgley (1978). *Beast and Man. the Roots of Human Nature*. Cornell University Press. Quoted from the London ed., Methuen, 1980.

M. F. Ashley Montagu (ed.) (1968). *Man and Aggression*. Oxford University Press.

M. F. Ashley Montagu (ed.) (1980). *Sociobiology Examined*. Oxford University Press.

Jesús Mosterín (1978): *Racionalidad y acción humana*. M., Alianza Editorial.

Javier Muguerza (1977): *La razón sin esperanza*. M. Taurus.

Ernest Nagel (1977): 'Teleology revisited', in *The Journal of Philosophy*, **74**, pp. 261–301.

Thomas Nagel (1973): 'Rawls on Justice', in *Philosophical Review*, **82**, pp. 220–234.

Thomas Nagel (1979): *Mortal Questions*. Cambridge University Press.

Friedrich Nietzsche (1881): *Morgenröthe*. Spanish ed., Aurora, trans. by P. González Blanco, Valencia, Sempere (2nd ed., B., Olañeta, 1978).

David Fate Norton and J. C. Stewart-Robertson (1980): 'Thomas Reid on Adam Smith's Theory of Morals'. In *Journal of the History of Ideas*, **41**, pp. 381–398.

Robert Nozick (1974). *Anarchy, State and Utopia*. Oxford, Basil Blackwell.

Peter O'Donald (1982): 'The concept of fitness in population genetics and sociobiology', in King's College Sociobiology Group (ed.), *Current Problems in Sociobiology*, pp. 65–85.

Maria Ossowska (1971): *Social Determinants of Moral Ideas*. London, Routledge & Kegan Paul.

Chaim Perelman (1976): *Logique juridique. Nouvelle rhétorique*. Paris, Dalloz.

Chaim Perelman (1977): *L'Empire rhétorique*. Paris, Vrin.

Kenneth Pike (1954): *Language in Relation to a Unified Theory of the Structure of Human Behaviour*, vol I, Glendal, Summer Institute of Linguistics.

Plato: *Obras completas*. Ed. de María Araújo and others. M., Aguilar, 1966.

Karl R. Popper (1965): 'Of Clouds and Clocks: The Arthur Holly Compton Memorial Lecture' C. W. Arensberg and H. W. Pearson (eds.), *Trade and Market in the Early Empires*, pp. 243–270.

Karl R. Popper, C. W. Asemberg and H. W. Pearson (eds.) (1957). *Trade and Market in the Early Empires*. N.Y., Free Press.

Karl R. Popper (1965): 'Of Clouds and Clocks: The Arthur Holly Compton Memorial Lecture'. Lecture delivered in Washington University, St. Louis. Reported in *Objective Knowledge* (1972), of which it forms Chapter 6.

Karl R. Popper (1972): *Objective Knowledge*. Oxford, Clarendon Press. Quoted from the Spanish ed., *Conocimiento Objetivo*, trans. by C. Solís, M., Taurus, 1974.

Karl R. Popper (1978): 'Natural selection and the emergence of mind', in *Dialectica*, **32**, pp. 339–355.

Karl R. Popper and John Eccles (1977): *The Self and Its Brain*. N.Y., Springer.

George E. Pugh (1978): *The Biological Origin of Human Values*. London, Routledge & Kegan Paul.

Hilary Putnam (1982): 'Why reason can't be naturalized', in *Synthese*, **52**, pp. 3–23.

Gerard Radnitzky (1983): 'The science of man: biological, mental and cultural evolution', in V. Cappelletti, B. Luiselli, G. Radnitzky and E. Urbani (eds.), *Saggi di storia del pensiero scientifico*, pp. 369–401.

D. D. Raphael (1928): 'Darwinism and ethics', in S. A. Barnett (ed.) *A Century of Darwin*, Chapter 8.

D. D. Raphael (ed.) (1969): *British Moralists*. Oxford, Clarendon Press (2 vols.).

D. D. Raphael (1980): *Justice and Liberty*. London, Athlone Press.

John Rawls (1971): *A Theory of Justice*. Harvard University Press. Quoted from the Oxford University Press ed., 1976.

Robert J. Richards (1982): 'Darwin and the biologizing of moral behaviour' in W. Woodward and M. Ash (eds.), *The Problematic Science*, pp. 43–64.

L. Robins (1932): *An Essay on the Nature and Significance of Economic Science*. London, Macmillan.

Francisco Rodríguez Adrados (1975): *La democracia ateniense*. M., Alianza Editorial. This is an abridged ed., adapted by M. Gonzalo, from *Ilustración y política en la Grecia clasica*, M., Revista de Occidente, 1966.

David Ross (1939): *Foundations of Ethics*. Oxford University Press. Quoted from the Spanish ed., *Fundamentos de ética*, trans. by Rivero and Pirk, Bs. As., Eudeba, 1972.

José Rubio Caracedo, Manuel Jiménez Redondo and Jesús Rodríguez Marín (1979). *Génesis y Desarrollo de lo moral*. Universidad de Valencia.

Michael Ruse (1979): *Sociobiology: Sense or Nonsense?* Dordrecht, Reidel.

V. Sánchez de Zavala (ed.) (1976): *Sobre el lenguaje de los antropoides* M., Siglo XXI.

John Schaar (1974): 'Reflections on Rawls' Theory of Justice', in *Social Theory and Practice*, **3**, pp. 75–100.

James Silverberg (1980): 'Sociobiology, the new synthesis? An anthropologist's perspective'. In G. W. Barlow and J. Silverberg (eds.) *Sociobiology: Beyond Nature/Nurture?*, pp. 25–74.

Peter Singer (1981): *The Expanding Circle. Ethics and Sociobiology*. Oxford, Clarendon Press.

J. J. C. Smart (1968): *Between Science and Philosophy*. N.Y., Random House, Quoted from the Spanish ed., *Entre ciencia y filosofia*, trans. by E. Guisàn, M., Tecnos, 1975.

Adam Smith (1759): *The Theory of Moral Sentiments*. (6th ed., amended in 1790). Ed. from A. G. West, N.Y., Arlington House, 1969. Quoted from the partial ed. of D. D. Raphael, *British Moralists (1969)*.

J. Maynard Smith (1972): *On Evolution*. Edinburgh University Press.

J. Maynard Smith (1976): 'Evolution and the theory of games', in *American Scientist*, **64**, pp. 41–45.

J. Maynard Smith (1982): 'The evolution of social behaviour – a classification of models', in King's College Sociobiology Group (ed.), *Current problems in Sociobiology*, pp. 29–44.

J. Maynard Smith and George R. Price (1973): 'The logic of animal conflicts', in *Nature*, **246**, pp. 15–18.

Elliott Sober (1981): 'The evolution of rationality', in *Synthese*, **46**, pp. 95–120.

Judy A. Stamps and Robert A. Metcalf (1980): 'Parent-offspring conflict', in G. W. Barlow and J. Silverberg (eds), *Sociobiology: Beyond Nature/Nurture?*, pp. 589–618.

Julian C. Stanley, William C. George and Cecilia H. Solano (eds.) (1977): *The gifted and the Creative: A Fifty-Year Perspective*. Baltimore, John Hopkins.

Charles L. Stevenson: *Ethics and Language* (1944). Yale University Press.

Avrum Stroll (1982): 'Primordial knowledge and rationality', in *Dialectica*, **36**, pp. 179–205.

P. W. Taylor (1961): *Normative Discourse*. Englewood Cliffs, N.J., Prentice-Hall.

René Thom (1968): Comments during the first Symposium (1966) of Villa Serbelloni on 'Ballard Mathews Memorial Lectures' of Waddington. In C.H. Waddington (ed.) *Towards a Theoretical Biology*, vol. I. *Prolegomena*.

Stephen E. Toulmin (1960): *An Examination of the Place of Reason in Ethics*. Cambridge University Press. Quoted from the Spanish ed., *El puesto de la razón en la ética*, trans. by I. F. Ariza, M., Alianza Editorial, 1979.

Stephen E. Toulmin (1981): 'Human adaptation', in U. J. Jensen and R. Harre (eds.). *The Philosophy of Evolution*, pp. 176–195.

Robert Trivers (1971): 'The evolution of reciprocal altruism', in *Quarterly Review of Biology*, **46**, pp. 35–37.

Ernest Tugendhat (1979): 'La pretensión absoluta de la moral y la experiencia histórica'. *Actas de las primeras jornadas de Etica e Ha. de la etica*. M., U.N.E.D. (in press).

Pieter A. Vroon (1980): *Intelligence on Myths and Measurement*. Amsterdam, North-Holland Publishing.

C. H. Waddington (1960): *The Ethical Animal*. London, George Allen and Unwin. Quoted from the Spanish ed., *El animal ético*, trans. by M. Marino, Bs. As., Eudeba, 1963.

C. H. Waddington (ed) (1968): *Towards a Theoretical Biology*. Vol I *Prolegomena*. Edinburgh University Press. Quoted from the Spanish edition (partial, *Hacia una biología teórica*, trans. de M. Franco, M., Alianza Editorial, 1976.

C. H. Waddington *et al*. (1942): *Science and Ethics* London, George Allen and Unwin.

James D. Wallace (1978): *Virtues and Vices*. Cornell University Press.

G. J. Warnock (1971): *The Object of Morality*. London, Methuen.

B. J. Williams (1980): 'Kin selection, fitness and cultural evolution' in G. W. Barlow and J. Silverberg (eds.), *Sociobiology: Beyond Nature/Nurture?*, pp. 573–588.

Edward O. Wilson (1975): *Sociobiology. The New Synthesis* Harvard University Press.

Edward O. Wilson (1978): *On Human Nature*. Harvard University Press.

Edward O. Wilson (1980): 'Comparative social theory' in Sterling McMurrin (ed), *The Tanner Lectures on Human Values*, vol I, pp. 51–73.

Edward O. Wilson and William H. Bossert (1971): *A Primer of Population Biology* Sunderland, Mass., Sinauer.

Edward O. Wilson (*vide*. Charles J. Lumsden and Edward O. Wilson, 1981).

Karl A. Wittfogel (1963): *Oriental Despotism*. Yale University Press. Quoted from the Spanish ed., *El despotismo oriental*, trans. by F. Presedo, M., Guadarrama, 1966.

Robert Paul Wolff (1977): *Understanding Rawls*, Princeton University Press.

William Woodward and Mitchell Ash (eds.) (1982): *The Problematic Science: Psychology in Nineteenth Century Thought* N.Y., Praeger.

V. C. Wynne-Edwards (1962): *Animal Behaviour in Relation to Social Behaviour* Edinburgh, Oliver and Boyd.

J. Z. Young (1978): *Programs of the Brain*. Oxford University Press.

INDEX OF NAMES

INDEX OF SUBJECTS